KAREN GARDINER

DESERT ESCAPES

· THE WORLD'S MOST INCREDIBLE PLACES TO STAY ·

Lannoo

INTRODUCTION

Desert escapes

I feel at home in the natural world but will always lean towards extreme landscapes, those places where nature makes you sit up and notice and that feel just a little bit more intense. And so I prefer my sand in a raw red desert than on a pristine white beach. A sunny beach is eager to please, but the desert challenges you to take it on. It's a forbidding landscape but one that is sublime in its solitude, starry skies, stillness and space.

My first taste of the desert was intoxicating. I was in Australia's Northern Territory where the hot air felt like a slap and the red earth looked like a firestorm. Having spent most of my life up until that moment on the northeast coast of Scotland, it was a sensation that felt utterly out of my then-small world. That sensation, and the desert, have tugged at me ever since.

The desert can be brutal, and without suitable shelter it will devour you. Building a shelter in an extreme landscape with limited resources is demanding, and yet, there's something about the challenge of a desert that inspires creativity. In the following pages you'll journey round deserts from the Gobi in Mongolia to the Atacama in Chile and discover accommodation as diverse as the landscapes. Some take their cues from their cultural environment, such as terracotta domes inspired by the traditional homes of the indigenous Himba of Namibia. Others take their inspiration from the surrounding flora-the exterior of a contemporary cabin that utilises the design of a tough endemic plant, for example. Many are concerned with making as gentle a footprint on the earth as is possible-such as light-touch tented villas that can be removed without a trace. And then there are those that evoke nostalgia for the past or the hope for a bright future, from renovated mid-century Airstreams to glass pods that look as if they've just touched down on a new planet.

While these escapes are far from simply places to lay your head and each is worth the trip alone, all are in locations that demand to be explored. In each of the following escapes, curious travellers will find opportunities to forge a deeper connection with their surroundings, whether by camel trekking over dunes, observing desert-adapted wildlife, or simply watching the sun dip below a sandstone arch.

09
AFRICA

87
ASIA, MIDDLE EAST AND EUROPE

CONTENTS

139
NORTH AND SOUTH AMERICA

241
AUSTRALIA

AFRICA

ZANNIER HOTELS SONOP

Old-world opulence in the middle of nowhere

Namib Desert

Sonop Farm, Road D707, Hardap Region

Namibia | +264 81 125 4930

zannierhotels.com/sonop

Deserts don't get much more desolate than the Namib in southwestern Africa. Believed to be the world's oldest desert, it's been arid or semi-arid for 55 to 80 million years. Stretching some 1,200 miles (2,000 kilometres) along Namibia's Atlantic coast and home to the highest sand dunes and some of the driest regions in the world, the Namib takes its name from the Nama word for 'immense', and it's easy to understand why. The landscape here is like nowhere else on earth and Zannier Hotels Sonop, at the southern end of the Namib, puts you right in the heart of it, with little as far as the eye can see but wide, open space and perhaps some desert wildlife such as oryx, leopards, hyenas, meerkats and jackals.

Approaching Zannier Hotels Sonop (by chartered flight or 4x4 vehicle) can make you feel like an intrepid ex-

plorer landing upon an undiscovered land. Indeed that's the intention: the property, set within 13,800 acres (5,600 hectares) of private, unspoiled desert, was designed to resemble a 20th-century tented camp and it wistfully evokes a bygone era of exploration and derring-do.

Constructed on top of granite boulders, amid soaring sand dunes, the hotel's ten spacious tented suites are elevated above the rocky desert floor by wooden stilts and decks and are covered with canvases that seem to melt into their surroundings. Each suite has several roll-up canvas openings so you can enjoy the surrounding scenery from multiple angles, including your four-poster bed and vintage soaking bath. Furnished with antiques and colourful rugs, the decor convincingly evokes the early decades of the last

century but modern touches such as air conditioning, Wi-Fi and rain showers keep things comfortable.

Beyond the rooms, the outdoor infinity pool offers superb views over the desert. The pool deck turns into an open-air cinema after dark and, after the film, guided stargazing sessions reveal the mysteries of the night sky -you can extend the evening with a cocktail or two at the gentlemen's club-style Cocktail & Cigar Lounge. Dinners in the main tent's candelabra-lit dining area are communal and you have five courses over which to swap tales with your fellow adventurers. Of course true explorers rarely sit still, so venture beyond the property in one of Sonop's guided excursions, from hiking and horseback riding (the lodge has its own stables) to taking to the skies in a hot-air balloon.

Lavish accommodation evokes a bygone era in the middle of nowhere.

Watch the desert change colours from your boulder-top tented suite. | Namibia | **Africa**

DESERT WHISPER

A place to embrace the solitude of the Namib Desert sands

Namib Desert

Gondwana Namib Park

| Namibia | +264 61 427 200 |

gondwana-collection.com

There's seclusion and then there's Desert Whisper, an intimate, one-of-a-kind desert hideaway for two. Resembling a spaceship that has touched down onto a rocky outcrop in one of the driest places on earth, this pod-shaped villa is one of just a handful of accommodation options in the private Gondwana Namib Park in the Namib Desert. Part of the sustainability-focused Gondwana Collection Namibia, it is perched above the desert sands in stark terrain about an hour's drive from Sossusvlei.

As otherworldly as Desert Whisper may appear, its eye-catching design was inspired by nature. Architect Sven Staby drew inspiration from the Namib's hardy endemic nara plant, which has a tough, leafless outer skin that can survive the harsh desert climate. This quality is reflected in the hard, curved metal exterior of Desert Whisper, which protects the soft, smooth interior where natural colours, materials and shapes similarly reflect the surrounding environment. Views from the enormous floor-to-ceiling windows take in the golden expanse of the Namib all the way to the horizon, fringed by the Naukluft Mountains in the distance.

Designed as a luxuriously intimate desert escape, Desert Whisper can only be booked for two people. You and your partner have exclusive use of the open-plan kitchen, dining and lounge area and plunge pool, which are connected to the villa by a snaking wooden walkway. Meals are prepared in advance according to guests' preferences and staff stay away as much as possible to ensure privacy (but are always within contact by radio). It's quite possible to spend days here encountering no other life but the passing oryx, jackal and ostrich, the only sound the soft whispers of the desert wind.

You could easily spend those days bouncing between the pool and terrace, refilling glasses of South African wines and gins from the fully stocked bar, but if you do feel like emerging from your desert hideaway, outdoor adventures abound. You can embark on a guided drive through the awe-inspiring Namib-Naukluft National Park, home to towering red sand dunes and the pans of Deadvlei and Sossusvlei, or rent an e-bike, grab a map and pedal your way towards the Naukluft Mountains.

Couples' getaways don't get more romantic than at this intimate escape in the midst of the Namib sands.

'Do Not Disturb' signs are superfluous at this secluded villa. | Namibia **| Africa**

WOLWEDANS COLLECTION

An enchanting desert retreat that gives back to the community

NamibRand Nature Reserve

NamibRand Nature Reserve C27	
Namibia	+264 61 230616
wolwedans.com	

In the Namib Desert, in the heart of the private NamibRand Nature Reserve, Wolwedans is so much more than simply a collection of timelessly elegant desert camps and villas. This is a place thoroughly committed to its founding ethos: to set an example in sustainability and fulfil its commitment to the conservation of the stunning nature reserve and development of its surrounding communities.

The Wolwedans Collection is comprised of three camps and two exclusive-use villas. Your options range from the four-room Plains Camp villa (the most exclusive one) to the down-to-earth, six-tent, Dune Camp. At the unique Desert Lodge, which has nine chalets, you can sleep just that little bit more soundly in the knowledge that you are helping to support the economic resilience of the community. Formerly known as Dunes Lodge, Desert Lodge

camp has been reinvented as the collection's hospitality training facility, part of the Wolwedans Desert Academy based at the village at Wolwedans. That means that the friendly young staff you encounter are getting practical, real-world experience in the hospitality business. Service might not always be 100 per cent seamless, but it will always be enthusiastic. It's all part of Wolwedans' vision for more inclusive conservation tourism that supports communities - an ethos any traveller with a conscience should easily be able to support.

Ultimately, though, you're here for the views. And Desert Lodge does not disappoint. Raised on platforms and stretching across a dune plateau, the chalets are all spaced apart to maximise privacy and designed so that bedrooms and bathrooms look out over the veranda to the desert and mountains-sleep

with your canvas blinds open for the soft sensation of the evening breeze and sight of the star-studded sky. A variety of communal recreation spaces offer plenty of options whether you're seeking solitude or ready to mingle: there's a spacious main area consisting of two lounges, several sundowner decks, a fireplace, a tea deck, a library, wine cellar and two dining rooms where you'll enjoy mouth-watering meals accompanied by excellent wines. The sparkling swimming pool is suspended above the dunes behind the lodge with 360-degree views.

Wherever you choose to spend your nights, daytime activities include bushman-led nature walks through the NamibRand Nature Reserve to learn about its desert-adapted flora and fauna; e-biking and horse-riding over the dunes, hot-air balloon rides; and scenic flights to the Diamond Coast.

Whether from the swimming pool or the sink, the desert views are endless. | Namibia | **Africa**

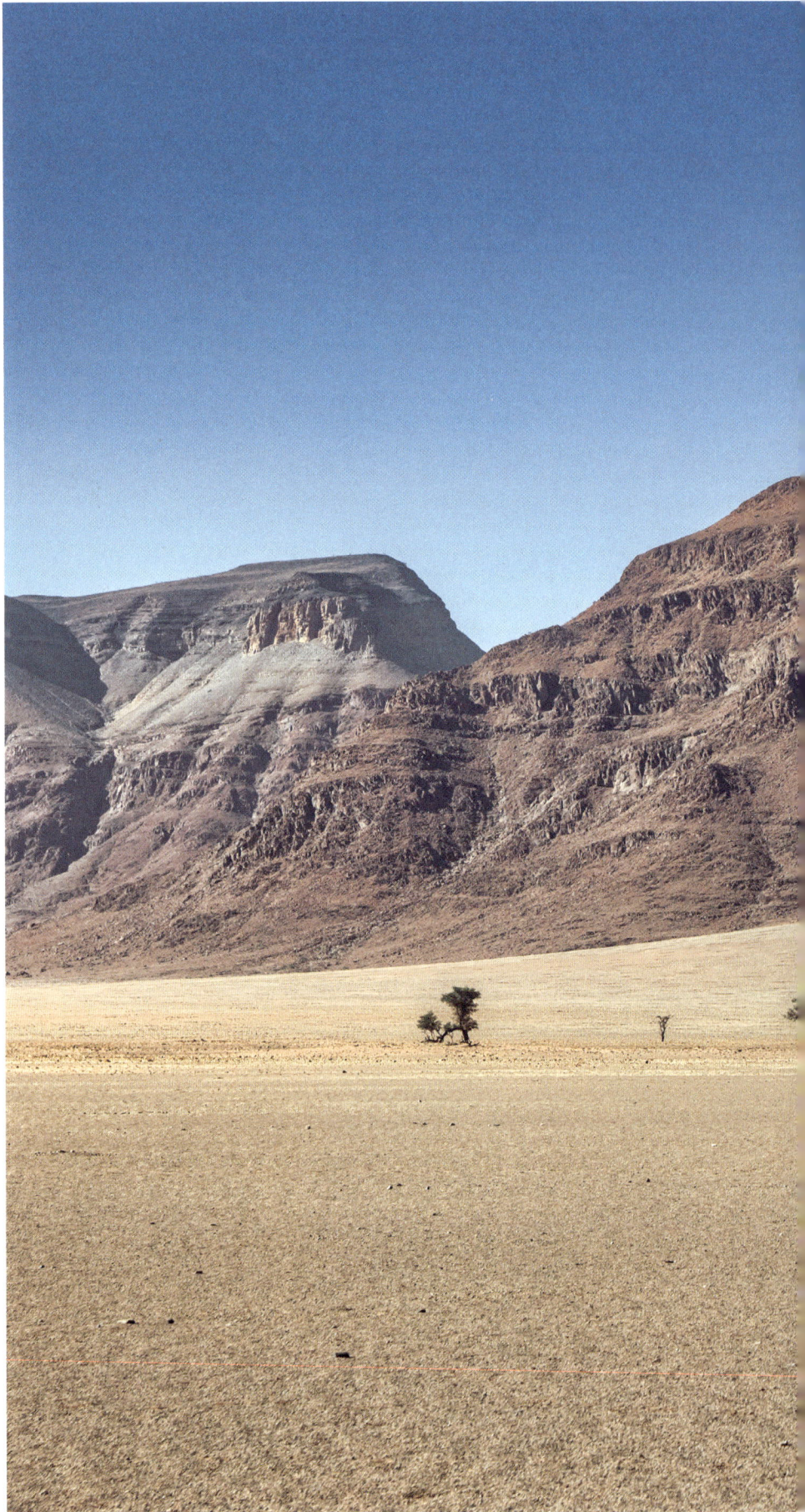

Enjoy the NamibRand's candy-floss colours from the comfort of your chalet.

LITTLE KULALA

Seclusion, sustainability and exclusive access to Sossusvlei

Namib Desert

Sesriem

Namibia | +264 61 274 500

wilderness-safaris.com

You can't get much closer to the iconic Sossusvlei sand dunes than Little Kulala. Situated in the 67,000-acre (27,000-hectare) Kulala Wilderness Reserve, in Namibia's hauntingly desolate Namib Desert, the lodge offers spectacular views of the famous burnt-orange dunes to the north; rugged mountains to the west, and vast open plains and the enigmatic grass-ringed formations known as 'fairy circles' in between. And with its own private entrance to Namib Naukluft Park, it's as secluded as it gets.

The entirely solar-powered camp's 11 spacious *kulala* lodges (meaning 'to sleep' in the indigenous Oshiwambo language) blend in effortlessly with their surroundings thanks to their soft, earthy desert tones and textures and use of local natural materials. Each has floor-to-ceiling windows from which to view the ever-shifting

landscape and spot passing desert-adapted wildlife such as ostriches, springbok and oryx, as well as the occasional brown hyena. Villas have private plunge pools and rooftop terraces that invite sundowners and stargazing or even (in the warmer months) a sleep out under the clear Namib sky. The elegant main lodge includes a library, wine cellar, craft boutique, dining area and lounge, and overlooks a watering hole popular with oryx, springbok and birdlife.

Little Kulala is operated by Wilderness Safaris, which runs dozens of luxury camps in remote, wildlife-rich wilderness areas across the African continent, each with an emphasis on sustainability. When the company arrived in the area about 25 years ago, many years of subsistence goat farming had left very little surviving indigenous wildlife. Today, with careful

rehabilitation, the land and its wildlife are back to their former glory and the Kulala Wilderness Reserve is nicknamed 'the living desert' because of its rich biodiversity.

See the desert come to life by embarking on an early morning excursion through the camp's exclusive entrance to the towering Sossusvlei dunes, including the famed 'Big Daddy', which stands just over 1,000 feet (325 metres) above the valley floor, making it one of the tallest dunes in the world. The camp offers guided flora and fauna walks along mountain and riverbed trails and scorpion-spotting night walks, as well as nature drives and electric bike and low-impact quad bike rides across the sand. There is also a seasonal sunrise hot-air balloon flight over the dunes, which ends with a delightful champagne breakfast.

Each suite has its own Star Bed so you can sleep under the stars and watch the night sky burst into life. | Namibia | **Africa**

OKAHIRONGO ELEPHANT LODGE

Rare wildlife and traditional culture draw adventurous visitors to one of Namibia's most remote lodges

Kaokoveld Desert

Purros Conservancy in the Kunene Streek, Purros	
Namibia	+264 61 237 294
okahirongolodge.com	

There's remote and then there's Kaokoland, a harsh, desolate, extreme, and dramatically beautiful region in the far north of Namibia that is considered one of Africa's few true remaining wildernesses. If you can make it there, you'll find one of the most original places you're ever likely to stay in: Okahirongo Elephant Lodge, in the Purros Conservancy, a 2.5-hour private flight or 12-hour 4x4 drive from Windhoek.

Perched on a hill above the banks of the Hoarusib River, at first glance the lodge, with its dramatic dome, resembles a James Bond villain's lair. The architecture of the lodge and its seven terracotta suites may appear unconventional, but they draw inspiration from the traditional homes of the semi-nomadic Himba, the region's indigenous people. And the design is not the most striking thing about the suites-save your astonishment for the views. From every angle, even while soaking in the huge bath, you get an eyeful of the Hoarusib River below and the jagged peaks of the Purros Valley Mountains in the distance. Turn your attention inside and the interiors won't disappoint either: each suite has four-poster beds and private terraces with enormous sun-loungers where you can sleep under the stars, and warm colours and textures throughout that echo the landscape. There's also one family suite with two bedrooms and the lodge has an infinity-edge swimming pool with views of the distant mountains. Meals are served in the open-air dining room and the menu is a fusion of Italian and African flavours prepared with ingredients that are mostly grown in the lodge's organic gardens. After dinner, you can gather round the fire and swap stories with the few other guests in the traditional boma.

But if you're staying at a place with 'elephant' in its name, you've probably come for the wildlife. Morning and afternoon game drives take you out in search of rare desert-adapted elephants, black rhino and lions, as well antelopes, giraffes, ostriches and zebras. The lodge also organises morning desert walks and visits to a nearby village where you can learn about the culture of the Himba. Staff can arrange a picnic lunch in the Hoarusib Valley after visiting the area's captivating 'clay castle' formations, and set up post-adventure sundowners on dunes overlooking the Skeleton Coast.

Africa | Namibia | Track desert elephants, rhino and lions through the remote Kaokoland.

SCARABEO CAMP

This secluded camp promises earthy luxury in a rough-hewn mineral landscape

Agafay Desert

Marrakesh	
Morocco	+212 6 62 37 62 95

scarabeo-camp.com

Think of a Moroccan desert and the rolling golden sand dunes of the Sahara are likely to spring to mind. But taking a trip to the Sahara demands a long drive from Marrakesh. Much easier and more accessible, albeit unconventional, is the nearby stone desert of Agafay, which lies at the foot of the snow-capped Atlas Mountains.

Just 22 miles (35 kilometres) from Marrakesh, but seemingly a world away from that city's frenetic energy, Scarabeo Camp offers the chance to sleep surrounded by the desolate grandeur of the Agafay Desert – only canvas between you and the star-studded sky-but without sacrificing the kind of elevated comforts you'd find in the luxury riads of town.

A collection of 15 white safari-style tents is dotted round the rough, lunar-like landscape, each positioned to maximise privacy and mountain views, and impressively decked out with plush and comfy beds, antique furniture and trinkets, and thick Berber rugs rolled out on floors. The camp's location changes twice a year: the owners adjust its position in consideration of seasonal conditions. In the summer months the camp sits in the cooling breeze and benefits from the shade provided by eucalyptus trees. The shorter, cooler days of winter see the tents positioned to soak up the full strength of the Moroccan sun.

Scarabeo Camp is a place where you could easily pass your days lounging around, enjoying the silence and serenity of the desert while sitting on your porch with a glass of sweet Moroccan mint tea in your hand. But when you're ready to leave your desert den, you'll find a full slate of activities on offer, each promising further immersion into the natural environment, from camel rides and quad biking to guided treks and early morning hot-air ballooning. As night falls, the desert casts a different spell. Tents are illuminated by flickering lanterns while guests dine on hearty Moroccan cuisine round candlelit tables. You can book a stargazing session with an astronomer for a deeper understanding of the star-spangled heavens. But no matter how deep into the night you stay up, the aroma of freshly baked bread and warm glow of the sun rising over the dusty desert will be sure to pull you out of bed the next morning.

Your stone desert oasis offers a front-row seat for pastel pink sunrises. | Namibia | **Africa**

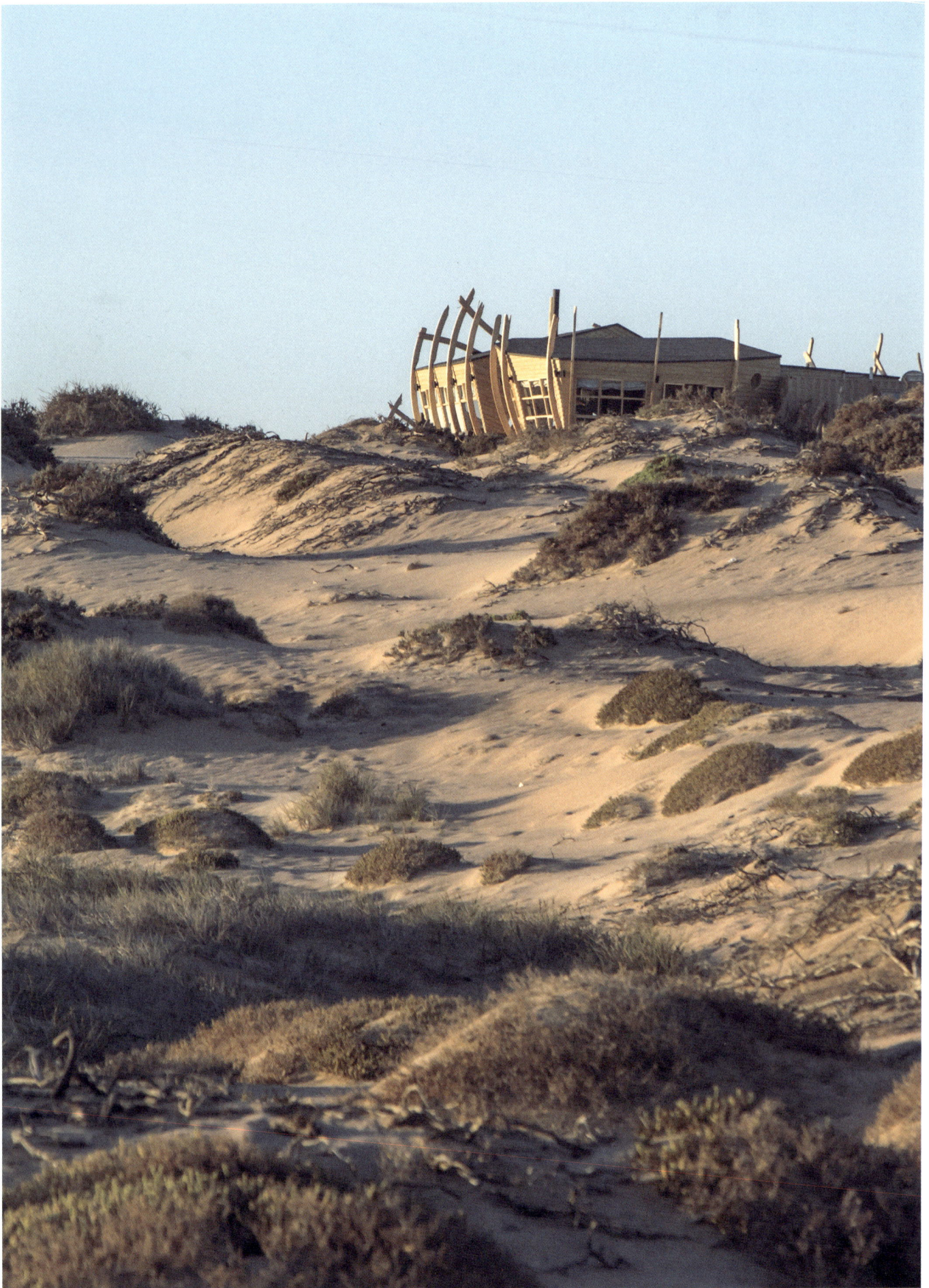

SHIPWRECK LODGE

Raw desert meets the Atlantic Ocean in this otherworldly outpost

Skeleton Coast

Skeleton Coast Park, Möwe Bay

| Namibia | +264 83 783 7055 |

shipwrecklodge.com.na

Shipwreck Lodge rises out of the sand dunes of Namibia's beautifully desolate Skeleton Coast National Park, a raw and forbidding place that takes its menacing name from the hundreds of ships that have met their end there over the past few centuries. Even today it may feel like the end of the world, its coastline scattered with ghostly, rusting wrecks; abandoned mines; and the huge bleached bones of long-departed whales. A walk on the beach here is like nowhere else on earth.

Located between the Hoarusib and Hoanib rivers, Shipwreck Lodge is the first and only permanent lodge in the park. Its angular wooden cabins face the Atlantic Ocean and their strikingly minimalist design gives them the impression of having quite literally washed up, broken and splintered, onto shore. But appearances here are deceiving. While it may look austere, Shipwreck Lodge's ten chalets are all warm and inviting places to ride out a sandstorm – or, less dramatically, a chilly desert night. The harshness of its wind-whipped environment is balanced by the comforts inside. All chalets are solar powered and feature a wood-burning stove and private deck. Glorious isolation is a major draw, but if you prefer company, you can mingle with fellow guests in the spacious, contemporary lounge and restaurant with a wraparound deck that offers sweeping views over the ocean –

ideal for watching the sun go down with a G&T in your hand. Shipwreck Lodge is part of the portfolio of Natural Selection, which owns a string of conservation-focused lodges across Africa, and it's an excellent base from which to explore the park – the lodge will happily arrange a tour for you. Roam the interior dunes and dry riverbeds to look for desert-dwelling lions, kudu and elephants; explore the craggy coastline in search of Cape fur seals lounging on the rocks or swimming in the foaming surf; look up and glimpse Rüppell's korhaans and Benguela long-billed larks in the sky; and pay homage to the fascinatingly dramatic history of this singular place by visiting the Suiderkus and Karimona shipwrecks.

The landscape is forbidding but Shipwreck Lodge's chalets are inviting. | Namibia | **Africa**

DAR AHLAM

Guests get the royal treatment in this 200-year-old kasbah

Ouarzazate

Skoura, Douar Oulad Cheik Ali

| Morocco | +212 524 85 22 39 |

darahlam.com

Hidden away at the end of a dusty track, just outside the small Moroccan town of Skoura, lies a labyrinthine, fortress-like, restored 19th-century kasbah. This is Dar Ahlam (or 'house of dreams', as its name translates), one of the most secluded and luxurious hotels in North Africa.

Dar Ahlam lies just south of the Atlas Mountains, on the fringes of Morocco's southern desert, and approaching it feels like catching sight of a welcoming oasis. Its high terracotta-coloured stone walls are shaded by a palm grove and surrounded by fragrant almond trees and carefully manicured gardens that were created by Louis Benech, a French landscape designer who updated Paris' Jardin des Tuileries and the Bosquet du Théâtre d'Eau at Versailles. Opulent fabrics and colours bring the surrounding environment

inside the 14 spacious suites, each of which has decadently large baths, two sitting rooms, an open fireplace and a plunge pool. Suites in the main building are dotted round a warren of scented corridors and alcoves and up hidden stairways, all of which intentionally creates an atmosphere that is extraordinarily discreet.

Service at Dar Ahlam is so pre-emptive that the staff seem to be mind readers. Meals, which fuse traditional Moroccan flavours with contemporary flair, are taken privately and when and where you choose, whether it be somewhere inside the kasbah or outside under the olive trees in the gardens. Long and luxurious days can be spent enjoying the large garden pool or the sumptuous hammam, Jacuzzi and massage rooms. But before getting too comfortable, make time to step out into the surrounding

desert wilderness on a personalised adventure with your private guide. You can embark on a camel ride across Sahara dunes, trek into the Valley of Roses (especially recommended in spring when the valley is in bloom), and drive to the hilltop vestiges of a 12th-century Berber village for a traditional tea ceremony at dusk. The hotel also offers the option to spend a night under the stars in the privacy of your own luxury tent.

Dar Ahlam is a long way from Marrakesh but another of the hotel's unique excursions allows you to make the journey the destination. 'The Memory Road', created by Dar Ahlam, is a magical seven-day journey from Marrakesh to Dar Ahlam which follows the oases of southern Morocco through landscapes that range from wild Atlantic coastline to the towering sand dunes of the Sahara.

Exceptionally well tended, the sprawling grounds promise sanctuary and solitude.

Sumptuous fabrics and rich Moroccon colours bring the environment to life. | Morocco | **Africa**

For an enchanting element of surprise, meals are set up in different locations each day.

UMNYA DESERT CAMP

Built by love, this intimate retreat is the perfect lovers' escape

Sahara Desert

Erg Chegaga, M'Hamid 47402	
Morocco	+212 600 66 66 16

umnyadesertcamp.com

Amid the rolling sand dunes of Erg Chigaga in the Sahara Desert, ten canvas tents draped in flowing white fabric form Umnya Desert Camp, an intimate hideaway built as an expression of the camp's founders' love for the desert and the nomadic lifestyle.

Each of the camp's ten tents is separated from the others by dunes ensuring seclusion, and raised above the desert sand by a wooden platform. You can choose between one of the superior suites, which are decorated with local crafts and have open living rooms, as well as private terraces where you can sleep under the stars; or the top-tier Royal Suite, which has the additional perk of more space and is set further away from the rest of the camp. If you're here for a romantic getaway, you can be sure of privacy whichever you choose as the camp can only accommodate 20 people at a time.

Immersion into the desert environment is at the heart of the Umnya Desert Camp experience. Activities on offer include sunrise or sunset camel rides, yoga, sandboarding, quad biking on the dunes, and visiting a traditional nomad settlement for lunch. It's evenings at the camp, however, that will create the most lasting memories: candles and lanterns are lit up round your tent to create an intimate atmosphere and you can cosy up next to the campfire and soak in authentic Berber culture through music and dance from folk musicians and belly dancers. Afterwards, discover the secrets of the clear, star-studded night sky during an astronomer-led stargazing session. Inside the largest tent you'll find the camp's restaurant, Lily Rose, which is focused on organic, vegan and authentic Moroccan cuisine, including dishes such as tagines and traditional soups and salads. Travelling couples also have the option of arranging a romantic picnic lunch or a dinner on the dunes accompanied by champagne.

The journey to the camp is an adventure in itself. The only way to reach its remote setting is by 4x4 drive – about three hours from Zagora Airport or eight hours from Marrakesh – so leave the driving up to the camp team who know the desert best: they will take you through the most scenic routes and, at your request, stop at interesting sites along the way. Summers are hot in the Sahara Desert – too hot – so the camp opens only from September to May.

Africa | Morocco | Relax amid rolling dunes at this secluded Sahara escape.

!XAUS LODGE

This community-run lodge offers the taste of an authentic way of life in a harsh landscape

Kalahari Desert

91st Dune, off the Auob River Road
Kgalagadi Transfrontier Park

South Africa | +27 (0)21 701 7860

xauslodge.co.za

Discover a safari lodge with a difference in the Kgalagadi Transfrontier Park, which straddles the South Africa-Botswana border. Owned by the ‡Khomani San (also known as Bushmen) and Mier communities, !Xaus Lodge lies deep in the Kalahari Desert and offers an experience that is as rich in culture as it is in wildlife.

Guests are picked up just outside Twee Rivieren and transported to the lodge in a 4x4 vehicle, along a soft sand track that crosses the Kalahari's rolling red sand dunes. Approaching the earth-coloured lodge, you'll struggle to pick it out of the landscape, so seamlessly does it merge into its natural surroundings. The complex consists of a central lodge and 12 individual chalets perched on a sand dune, each with its own private deck overlooking a vast

heart-shaped salt pan (!Xaus means 'heart' in the Nama language) with a freshwater hole, which attracts frequent four-legged visitors. Rustic furnishings and eye-catching artworks throughout the lodge are all made by local artisans, whom you can watch at work in the lodge's recreated cultural village.

Activities on offer include guided early-morning walks through the dunes with San trackers and sunset game drives in an open safari vehicle in search of black-maned lions, leopards, gemsbok, meerkats and more than 260 species of bird, including two-thirds of the raptor species found in southern Africa. Nighttime brings clear skies, bright stars and the opportunity to listen as Bushmen tell you their traditional legends of the night sky. The !Ae!Hai Kalahari Heritage Park

that surrounds !Xaus Lodge is an International Wk Sky Sanctuary and offers some of the most dazzling stargazing opportunities in South Africa. The lodge's sparkling swimming pool is the perfect spot for cooling off in the afternoon sunshine and the central area's bar, lounge and dining area offer warm hospitality and hearty cuisine.

While there's much to do, the real beauty of !Xaus Lodge is that it offers a rare opportunity to slow down, switch off and allow yourself to be immersed in the peace and tranquillity of its desert environment. While service is top-notch, its remote location means that harsh conditions prevail; the power supply is limited and there's no mobile or Wi-Fi coverage, though that absence may be welcomed by those seeking to detox from the modern world.

!Xaus Lodge offers some of the best star gazing in South Africa. | South Africa **| Africa**

SANBONA WILDLIFE RESERVE

A Big Five wilderness reserve within reach of the big city

Klein Karoo

R62, Montagu, 6720

South Africa | +27 (0)21 010 0028

sanbona.com

Just a three-and-a-half-hour drive along the scenic Route 62 from Cape Town, Sanbona offers the chance to escape the hustle and bustle of the city and enjoy the serenity of a raw landscape of undulating mountains and plains. The private reserve lies at the foot of the towering Warmwaterberg mountain range in the heart of the stark Klein Karoo, a semi-desert region in the Western Cape. Steeped in indigenous San culture and heritage – there are seven recorded rock art sites throughout Sanbona that date back more than 3,500 years – and home to rich flora and fauna, including the Big Five (lion, leopard, black rhinoceros, elephant

and buffalo), Sanbona is a hidden gem that shouldn't be overlooked.

Sanbona contains four different accommodation options, each with its own character. Dwyka Tented Lodge is made up of nine tented suites, each with a private deck and plunge pool overlooking the open plains; Tilney Manor's six suites promise a stately, exclusive oasis with mountain views and all the creature comforts, and the 12-suite Gondwana Family Lodge is a fun and relaxed choice for those travelling with younger ones. The Explorer Camp, where wild animals may pass by the three canvas tents at any given time, is the top pick for the more adventurous travellers.

Sunrise and sunset nature drives, as well as guided nature walks, are included in rates. While travelling the reserve in search of wildlife, you'll journey through two different biomes and several habitats, including the quartz fields of the Karoo, which is characterised by beds of angular quartz debris and home to tiny succulent shrubs. As you explore, guides will share the story of how Sanbona serves as a major role player in the conservation realm. By rehabilitating land that had been transformed by intensive agriculture, and reintroducing endangered wildlife species, Sanbona Wildlife Reserve is a true example of a successful rewilding project on an epic scale within the Western Cape.

Discover the Big Five in Sanbona's private reserve. | South Africa | **Africa**

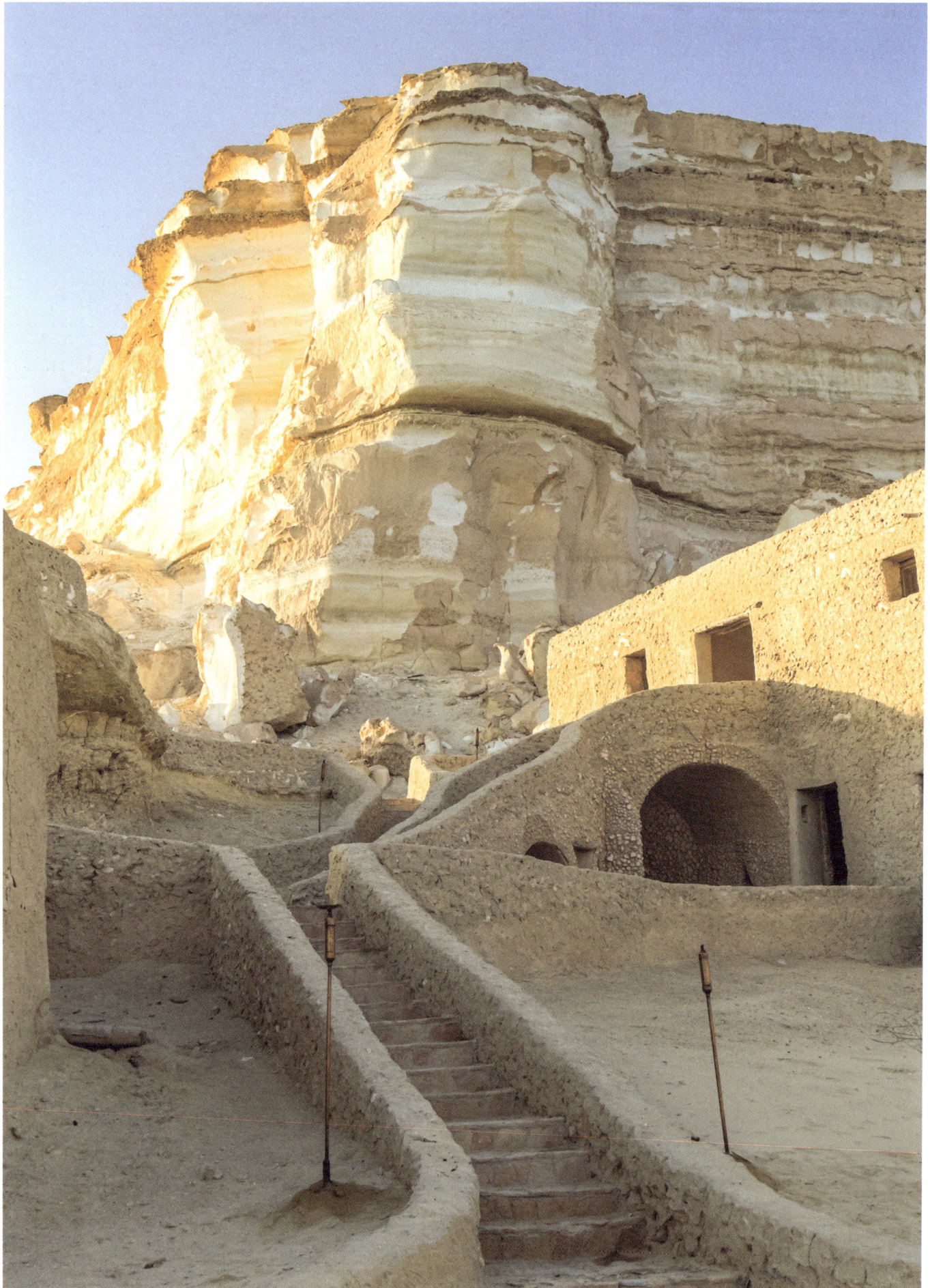

ADRÈRE AMELLAL

This sandcastle-esque escape in Egypt's Western Desert is an oasis within an oasis

Siwa Oasis

Gaafar Mountain, Siwa, Matrouh Governorate

Egypt | +20 22736 7879

adrereamellal.com

Backed by the massive limestone White Mountain (*Adrère Amellal* in the Berber language spoken here) and rising over Siwa Lake like a giant sandcastle, Adrère Amellal can surely claim to be one of the most unusual hotels in the world.

Built using *kershef* (salt rock and clay), a material used in the Siwa Oasis for centuries, the earth-coloured hotel blends naturally into its environment - so much so that, approaching this whimsical desert escape with its curving lines and soft turrets, situated so far west of Cairo it's almost in Libya, you may feel like an ancient mariner, tricked by a *fata morgana* into seeing fairy castles in the air. But as improbable as it may seem, it's real. Rub your eyes and it's still there: a cluster of playfully shaped buildings, some

square, some rounded, house 40 rooms where the *kershef* walls absorb the searing desert heat to keep temperatures down during the day, then radiate warmth at night when the temperature drops. The simple interiors feature chairs made of palm leaves, rugs woven by local artisans, and beds and night tables carved out of salt rock. Roofs are covered with palm leafs and windows are thoughtfully sized and placed to catch desert breezes. There's no need for air conditioning here, which is just as well because, with an eye to minimising its impact on the environment, Adrère Amellal has no electricity. At night, the property relies on starlight and beeswax candles for light. Water comes from the more than 200 springs in the oasis, the lodge garden provides organic food for guests' meals, and

a natural hot spring serves as a swimming pool where guests can take a dip amid the palm groves.

If you want to use your mobile phone at Adrère Amellal, you are only permitted to do so in your own room. But why would you? The freedom from technology and the trappings of contemporary life offered here gives a whole new meaning to the idea of a luxurious escape. Leave your phone behind and immerse yourself in your surroundings. Go sandboarding on the Great Sand Sea or horse riding through desert dunes, or explore the great historic sites of Siwa. Its ancient temples and fortresses have architectural techniques that echo through the centuries and find modern expression in the extraordinary lodge.

This clay- sand- and salt-built shelter promises a minimalist yet luxurious stay. | South Africa | **Africa**

SAN CAMP

The epitome of the classic romantic African hideaway

Kalahari Desert

Makgadikgadi Pans National Park, Gucta

Botswana | +27 11 326 4407

naturalselection.travel/camps/san-camp

Seven white billowing canvas tents are perched amid grassland on the edge of the shimmering Ntwetwe Pan in Botswana's legendary Makgadikgadi. One of the largest salt pans in the world, Makgadikgadi was once a super lake covering most of Botswana; today it is a place of ethereal beauty, magic and mystery. San Camp is an oasis in this spectacularly stark environment and the views from the camp stretch all the way to the horizon, interrupted only by scattered desert palms and baobabs.

Inside those light-filled tents you'll find enormous four-poster beds raised high above Persian rugs, mahogany writing desks and leather armchairs – an elegant contrast to your rugged surroundings. Part of the conservation-focused Natural

Selection portfolio of owner-operated safari camps, the entire camp runs on solar power and leaves little mark on its environment. When darkness falls, you'll find lanterns scattered round the camp to light the way and set the mood.

In the main area, bright and breezy open-sided pavilions house a mess tent where you'll enjoy long, decadent meals between game drives a peaceful yoga and meditation pavilion where you can let your mind and body unwind, and a tea tent with Persian carpet cushions where you can swap stories with fellow travellers. The views command the attention, but don't overlook the camp's Natural History Museum cabinets, which are filled with museum-worthy collections of old maps, fossils and artefacts.

Open only in the dry season, from early April to mid-October (the only time that the Ntwetwe Pan is accessible), San Camp offers multiple special ways to explore the Kalahari. You can search for wildlife such as lions, wildebeest, zebras, elephants and even rare brown hyenas by taking a game drive in a 4x4 that has been custom built to tackle the challenging terrain, and enjoy up-close encounters with cheeky meerkats. You can join Zu/'hoasi Bushmen on bush walks to learn about how the indigenous people have long survived and thrived in such a seemingly inhospitable environment. Adrenaline-seekers can speed across the salt-crusted earth by quad-bike, then relax by lying out on the pans as the sun sets and a planetarium of stars appears overhead.

Billowing white tents and a mysterious desert landscape create the setting for a romantic retreat.

It has provided the backdrop for fashion shoots and music videos, but the camp steals the show. | Botswana | **Africa**

ASIA, MIDDLE EAST AND EUROPE

TELAL RESORT AL AIN

This serene desert resort offers the perfect blend of natural wonders and cultural heritage

Remah Desert

Remah, Al Ain, Abu Dhabi

United Arab Emirates | +971 3 702 0000

telalresort.ae

An Arabian oasis surrounded by towering caramel-hued sand dunes, Telal Resort is set in the heart of the vast Remah Desert. About an hour from Al Ain and two hours from Abu Dhabi, the resort makes for an easy escape from the hustle and bustle of the city and overlooks a natural conservation area home to rare plants and roaming desert gazelles, antelopes, reem and Arabian oryx.

Telal Resort is comprised of a main building with a restaurant, library, spa and lagoon-style infinity pool, as well as 24 traditionally decorated rooms and villas spread out among the surrounding dunes. Accommodation options run the gamut from entry-level Heritage Rooms to spacious Gazelle private villas with bespoke furniture, custom-made chandeliers and hand-carved baths; the tented Domani two-bedroom Pool Villa with open terrace and a private plunge-pool; the romantic Reem Presidential Pool Villa set in a secluded sand dune; and the top-tier Arabian Oryx Royal Pool Villa, also set on its own dune, which has four large bedrooms, a private pool, sauna, steam room and Jacuzzi.

Ten detached suites are situated at the Zaman Lawal Heritage Village, next to the main building and home to a souk, traditional Arabian restaurant and other cultural attractions enabling guests to explore the rich Bedouin heritage of the Al Ain region. All rooms offer complete privacy, uninterrupted views of the ever-changing colours of the landscape and direct desert access so you can step barefoot onto the sand right outside your door and head out for a solitary stroll across the curving spines of the giant dunes. On-site activities include falconry displays, sand biking, horse and camel riding, desert wildlife safaris and traditional cooking lessons. Meals are served with Middle-Eastern flair in the Desert Gate restaurant and guests also have the romantic option of a private dinner in a candlelit dune-top cabana.

If you want to dive deeper into the cultural heritage of the region, make the short trip to Al Ain Oasis. A central element of the UNESCO Cultural Sites of Al Ain, the oasis is fed by a traditional *falaj* irrigation system and preserves an ancient way of life. Farmers tend to more than 147,000 date palms here, as well as fodder crops and fruit trees such as mango, orange, banana, fig and jujube.

Rejuvenate body and mind in the peaceful surroundings of the Remah Desert.

Middle East | United Arab Emirates | Dine on the dunes for a sumptuous Arabian spread with a view.

HABITAS ALULA

*Cultural heritage and contemporary art transform
an ancient desert valley into an emerging destination*

Ashar Valley

AlUla

Saudi Arabia | +966 14 821 3900

ourhabitas.com

Amid desert cliffs of sandstone in northwest Saudi Arabia, AlUla is one of the oldest regions in the Arabian peninsula. A central trading crossroads along the Silk Road, as well as the Incense Route from southern Arabia, AlUla is steeped in many thousands of years of human history. Now, travel to the kingdom having eased over the past few years, the region is poised to be rediscovered by modern travellers.

Leading the way in the region's tourism resurgence is Habitas, the experience-led hospitality brand known for its focus on sustainability, wellness, adventure, music, food and culture. The brand's first Saudi outpost, Habitas AlUla, is unobtrusively nestled within an ancient oasis in the desert canyons of the Ashar Valley and surrounded by ochre cliffs and palm groves. The Habitas model is conscious of its impact on its surroundings and so all building elements are constructed using an innovative modular build, then flat-packed and installed so as not to leave a trace on the environment. The 96 stand-alone villas are generously spaced out to ensure maximum privacy and views of the golden sands and rocky, wind-sculpted outcrops stretching to the horizon. Sister property Caravan AlUla, a few minutes' drive away, offers an immersive glamping experience for adventurous travellers and is made up of 22 adapted Airstreams, a large common lounge tent that encourages guests to mingle, and a cluster of food trucks.

Meals at Habitas AlUla's restaurant Tama, which means 'here and now' in Aramaic, fuse global and Middle-Eastern cuisine and bring Saudi Arabia's unique flavours and heritage to life by using local, fresh ingredients and spices that were traded along the Incense Route. There's also a spa, fitness centre and swimming pool, all connected by pathways that meander through the property and can be traversed by foot or by provided e-bikes and golf carts.

Guests are invited to widen their experience of this historical region, which is something of an open-air museum, by taking guided walks through the remains of the ancient city where the Nabateans once lived. Thousands of tombs and burial places can be found in AlUla, including more than 100 intricately carved tomb facades at Hegra, Saudi Arabia's first UNESCO World Heritage Site. Other possible adventures include leisurely camel rides and adrenaline-fuelled dune buggy rides through sandstone mountains, into narrow canyons, and over sand dunes.

While cultural heritage is AlUla's most obvious draw, it also boasts an impressive contemporary art scene, thanks to the biannual Desert X AlUla, which invites artists to engage with the desert environment to create site-specific works, several of which remain in place beyond the biennial.

Contemporary artworks amplify the beauty of the desert.

Villa design pays tribute to the craftsmanship of Saudi Arabia. | Saudi Arabia | **Middle East**

SIX SENSES SHAHARUT

Detox, decompress, and dive into the Negev Desert's ancient heritage

Negev Desert

Hevel-Eilot, Shaharut

Israel | +972 86 15 00 50

sixsenses.com

Off a dusty road in Israel's Negev Desert, amid bluffs and dunes, a simple gate marks the entrance to Six Senses Shaharut. The subtlety of the welcome gesture sets the tone for the experience. This may be a luxurious spa retreat, but it doesn't need to shout about it. Low-slung, softly curved buildings made of stone quarried from nearby valleys peek modestly out from atop a high ridge, blending seamlessly with their surroundings.

Tel-Avid-based Plesner Architects took design inspiration from the architecture of the Nabataeans, an ancient nomadic Bedouin tribe. The resort's 60 suites and villas, spa, pools and restaurant were carved out of the mountain overlooking the Arava Valley. The aim was to create unobtrusive buildings that look as if they've been there since the long-ago days when the tribe roamed the Arabian Desert. The generously sized accommodation invites the landscape inside with floor-to-ceiling windows

offering panoramic views of the Negev Desert and Arava Valley as well as the ever-changing colours of the Edom Mountains in the distance.

Helping guests to connect with themselves is as much a feature of Six Senses' luxury hotels as connecting with the surrounding environment, and Shaharut is no exception. As you'd expect from a brand much loved for its immersive spa experiences, the wellness programme here is impressive. The signature spa has six rooms for individual treatments such as massages, facials, and wraps using local products including olive oil, camel milk and herbs from the hotel garden, and also offers more holistic, multi-day healing programmes to rejuvenate, detox and improve sleep. There are also saunas, steam rooms and hammams, as well as a large indoor pool, outdoor lap pool, gym and yoga studio with desert views. From early-morning camel trekking

across the Negev dunes to following ancient trails along the historic Incense Route by Jeep, a suite of activities will tempt you out of your luxurious suite or villa each morning. The resort can arrange a visit to the local kibbutz and guests are also welcome to visit the resort's own camel stables to learn about how they groom, feed and care for the animals. Afterwards, choose from several dining venues for dinner, each combining Israeli and Mediterranean influences and, where possible, using ingredients harvested from the resort's organic gardens and nearby kibbutzim, then discover some of the clearest skies in Israel during a guided stargazing session.

Your adventure to Six Senses Shaharut begins with a transfer from Tel Aviv or Jerusalem, which can mean travelling through biblical, dusty landscapes by road or chartering a helicopter for the ultimate bird's eye view.

Suites and villas are set amid the bluffs and dunes of the Negev Desert.

Feel stress melt away in your outdoor tub or private pool. | Israel | **Middle East**

AL MAHA DESERT RESORT AND SPA

Forget what you think you know about Dubai at this secluded desert oasis

Dubai Desert

Dubai Desert Conservation Reserve,
Al Ain Road, Dubai

United Arab Emirates	+971 4-8329900

marriott.co.uk

If it's a thoroughly indulgent luxury desert retreat you're looking for, Al Maha Desert Resort and Spa is just the place. A contemporary interpretation of a Bedouin village, the resort is tucked away in a palm grove-dotted oasis where 42 lavish tented suites boast private infinity plunge pools and terraces with views of rolling sand dunes from every angle. Suites are generously spaced out to ensure maximum privacy and are furnished with elegant handcrafted Arabian antiques and artefacts.

Situated in the protected Dubai Desert Conservation Reserve, Al Maha is just an hour's drive from Dubai but feels a world apart: here the gnarled traffic, overdevelopment and shimmering skyscrapers of the frenetic city are replaced by free-roaming gazelles, oryx, camels, and endless vistas of the sweeping plains and Al-Hajar mountain range – don't be surprised to spot a passing animal or one of the region's 130 species of bird taking a sip from your pool or one of the many watering holes dotted round the grounds.

Complimentary excursions and activities offer the chance to fully immerse yourself in the desert environment and may include falconry, archery, camel trekking and horseback riding. You may take a wildlife-focused 4x4 drive over the sand dunes for up-close views of the once-rare, now-plentiful Arabian oryx after which the resort is named (*al maha* means Arabian oryx in Arabic) as well as other indigenous fauna and flora in their natural habitat. Activities are led by enthusiastic naturalists and conservationists who are employed by the resort to share the area's natural heritage and culture with guests.

When not out exploring the desert, guests can visit the Timeless Spa for an Arabian-inspired *rasul* chamber experience (it's a skin smoothing treatment involving mineral-rich mud and steam). Dining is delightful at Al Diwaan restaurant, where the menu blends local and international flavours and every table offers spectacular views. If you want to amp up the romantic factor, book the resort's intimate Dune Dining experience and enjoy a lantern-lit dinner in a secluded setting under the twinkling stars.

From camel trekking to falconry, Al Maha's included excursions will dazzle even the most well-travelled visitors. | United Arab Emirates | **Middle East**

THREE CAMEL LODGE

This one-of-a-kind lodge captures Mongolia's nomadic spirit

Gobi Desert

Dalanzadgad

Mongolia | +976 11 325 786

threecamellodge.com

Against a mountainous backdrop, felt- and canvas-covered gers are scattered across a wild expanse of the Gobi Desert. Mongolia's nomadic herders have used tents just like these for millennia, assembling and disassembling them as they follow their flocks. Although Three Camel Lodge's collection of gers offer a few more home comforts than is customary, they keep closely to tradition and are all locally made using native techniques and materials. That's just one part of this astonishing eco-lodge's commitment to celebrating Mongolia's cultural traditions and preserving its spectacular environment.

Three Camel Lodge's 40 gers are each made with a wooden frame of fanned-out poles resting on latticed wooden walls. They are all beautifully appointed with furniture that has been hand- carved and painted by local artisans and made warm and welcoming with thick wool carpets, chunky camel-hair blankets and wood-burning stoves. The gers and main lodge were all designed in the style of ancient Buddhist temples and built using traditional wooden pegs with not a single nail in the roofs. Roof tiles were moulded by hand in a centuries-old technique while electricity meets modern needs and is generated by solar and wind power. Each ger's door faces south for views of the vast desert plains and the Gobi-Altai mountain range.

Three Camel Lodge is an excellent base for exploring the Gobi with its vastly different ecosystems, rich wildlife, and fascinating sites of historical and natural significance; staff will assist with all arrangements. The red sandstone Flaming Cliffs are a must-see, particularly at sunset, when they are cast with a magical glow. There's also Yol Valley National Park, a deep desert valley that is surprisingly green and the habitat of vulture-like lammergeiers, Altai snowcocks, ibex, yaks and argali mountain sheep. Riding Bactrian camels across Moltsog Els, a magnificent array of sand dunes, in the company of a local herding family, and viewing petroglyphs etched by early Gobi settlers more than 5,000 years ago are just two more of the endless culture-focused adventure options on offer here.

After a day of exploring, a Mongolian-inspired massage revitalises body and spirit before refuelling with seasonal fare at the Bulagtai Restaurant, which uses locally grown organic produce and organic meat from Mongolia's free-range livestock herds. Afterwards, enjoy handcrafted cocktails and traditional Mongolian dance and throat-singing in the Dino Hall and end the night with a stargazing session back at your ger.

Cultural preservation and environmental stewardship are at the heart of this Mongolian eco-lodge. | Mongolia | **Asia**

SUJÁN
THE SERAI

A blissfully serene wellness destination
near the "Golden City" of Jaisalmer, Rajasthan

Thar Desert

Rajasthan

India | 91 11 4617 2700

thesujanlife.com

Established by two passionate conservationists, Jaisal and Anjali Singh, the Suján brand is committed to preserving, protecting, and restoring the Indian wilderness, and includes three exclusive tented camps in Rajasthan. For those who wish to discover the stark beauty and diverse cultural heritage of the Thar Desert, a region of rolling sand dunes in the northwest of the Indian subcontinent, Suján The Serai makes the perfect basecamp.

Set on a 100-acre (40-hectare) private estate of indigenous desert scrub, the camp is a tranquil oasis less than an hour's drive from Jaisalmer. The Serai honours the cultural heritage of its environment by drawing inspiration from the designs of the royal caravan sites of the Rajput period while maximising comfort with modern amenities and unobtrusive service. The camp's 21 white canvas tents are raised on plinths of honeyed Jaisalmer sandstone and each has its own outdoor veranda. Six luxury tented suites have their own private walled-in gardens and sunken, heated plunge pools with Jacuzzi jets, while the ultra-exclusive Royal Tented Suite has its own spa, outdoor pool, and dining and lounge tents walled into a private encampment.

Wherever you spend your nights, your days will be filled with camel safaris, birdwatching expeditions, desert drives, or jaunts to Jaisalmer's fabled golden fortress, a UNESCO World Heritage Site. Beyond the landscapes and history, the Serai is keen to introduce guests to the region's living traditions. You can watch artisans at work on the camp's sand deck spinning pottery or carving wood, and listen to the hauntingly beautiful melodies of traditional Manganiyar folk musicians. Then again you could simply embrace the slow pace of desert life by retreating to the spa, where treatments are offered in breezy tents within a walled garden.

Sundowners on the dunes are followed by exceptional dinners, prepared using produce from the Serai's own organic garden. Choose to dine inside the dining tent, on the stone deck, by the large pool, or privately with a lantern-lit dinner under the star-studded sky, accompanied by local folk musicians.

Because the searing desert heat of the Thar Desert gets extreme in summer, the Serai is open only from September to April.

The Serai is a serene retreat surrounded by nature.

The wilderness spa creates a sense of wellbeing with its array of soothing treatments. | India | **Asia**

KANER RETREAT

India's first desert botanical resort blossoms in Rajasthan

Thar Desert

Jodhpur-Jaisalmer Highway,
Village Dera, Jodhpur

India | +91 98102 03057

kanerretreat.com

Within just two hours' reach of the blue city of Jodhpur, but seemingly much further from the clamour of urban life, Kaner Retreat provides an intimate introduction to the beauty and botanical bounty of northern India's Thar Desert.

Taking its name from the oleander flower, also known as "desert rose", Kaner Retreat has ten light-filled villas. Each is themed round a local flower, reflected in specially commissioned botanical art and the colours of textiles and furnishings that incorporate local materials and reclaimed wood sourced from handicrafts factories in Jodhpur. Antiques have been thoughtfully sourced locally to enhance the rooms' eclectic charms, and private courtyards provide a peaceful setting to relax among fragrant flowers. Red stone exteriors act as a shield in summer and suntrap in winter. Round the resort there are walking paths, a glasshouse lounge with 360-degree desert views, and an intimate, stepwell-inspired pool. India's first desert botanical resort was brought to life by Sapna Bhatia, who grew up in the Thar Desert. After many years working as a journalist in Delhi and London, she drew upon her love of her childhood landscapes and opened the hotel to share its magic with guests. Surrounded by an *oran*, or sacred grove, Kaner Retreat sits in an unspoiled environment where local animals are free to graze but no agricultural practices, such as pruning and cutting trees, are allowed.

The landscapes of this place feed its cultural traditions – the area's literary, musical and dance traditions all draw inspiration from the desert. Guests can find their own inspiration by accompanying Bhatia on a guided botanical walk in the sacred grove and Kaner's own micro nabitats. They may embark on an exhilarating jeep safari (or slower-paced camel expedition) into the surrounding Thar Desert.

Horses, too, are an integral part of Thar culture and guests have the opportunity to ride and learn about them with a visit to a local stud farm. At the nearby Khichan Bird Sanctuary birders can see elegant demoiselle cranes, which return from Siberia to Rajasthan every March. Some say the birds make the arduous journey for Rajasthan's famous hospitality – it's not such a far-fetched notion: villagers have created a *chugaghar* (feeding place) for the birds.

Customisable dining experiences include starlit dinner on private sand dunes and lunch in an olive grove. Whichever experience you choose, you won't go hungry: meals range from modern Rajasthani five-course dinners to traditional thalis made from revived traditional recipes and incorporating ingredients foraged nearby.

Kaner Retreat operates seasonally, from October to March.

From culinary traditions to botany, this retreat offers an introduction to life in the Thar Desert. | India | **Asia**

THOUSAND NIGHTS

*Find silence and splendour amid
the rippling dunes of the Sharqiya Sands*

Sharqiya Sands

Madinat Al Sultan Qaboos

Oman	+968 99448158

thousandnightsoman.com

Set amid undulating desert dunes, in a valley in the heart of Sharqiya Sands in eastern Oman, Thousand Nights is named after the number of days that it took its founder, the traveller and writer Abdullah Akhdar, to come upon the desert escape of his dreams. From a young age, Abdullah nurtured a deep love of the desert's shimmering sands, the dance of the dunes, and songs of the wind. After an unsatisfying stint of living amid the hustle and bustle of the big city, he embarked on a quest to find the place of his dreams, where he could experience solitude and wonder under a blanket of stars. He searched for 999 nights. In 2001, on the thousandth night, he arrived at this tranquil oasis where green trees dot a landscape of brilliant reds and bright golds, creating a mesmerising contrast between bright sands and cooling shadows. Deciding that such a place could not be kept secret, he created the Thousand Nights camp to share some of its magic.

The Bedouin-style resort invites seekers of tranquillity and a touch of adventure to soak up the magic of the Arabian Desert. Made from goat's wool, decorative and colourful tents are scattered round the camp and furnished with Arabic rugs. There's also the option of upgrading to a glass-walled luxury Ameer Tent or to a Sand House, a small but spacious villa with an open balcony. Desert pursuits such as camel trekking in the dunes, sandboarding and dune bashing can all be arranged on the property.

Given that it is deep in the desert, Thousand Nights surprises with the extent of its facilities – there's even a small swimming pool to provide respite from the heat of the day. Designed to resemble an Omani fort, the main building houses the restaurant and lounge area where you can wake up with a hearty Bedouin breakfast, unwind with an authentic Arabian coffee or enjoy pre-dinner drinks. In the evening kerosene lamps are lit along the pathway to the dining area where Omani favourites, such as *shua* (slow-roasted lamb), are served round a crackling bonfire while Bedouin musicians play traditional songs.

Scramble to the top of the sand dunes for dramatic sunset views. | Oman | **Middle East**

THE COLONY PODTEL

This 100-per-cent self-sufficient pod promises wonderful isolation in a surprising location

Gorafe Desert

Gorafe

Spain | +351 22 145 0770

districthive.com

The Gorafe Desert, in the Andalusian province of Granada, is a landscape of burnt-orange canyons and deep gorges that can give you the impression of having arrived in the American West rather than Western Europe. So surreal is this baked and beguiling environment that you may not think twice about coming across a sleek capsule hotel hovering slightly above the ground as if it were an alien spacecraft contemplating making first contact.

Set against the stunning backdrop of the Sierra Nevada, The Colony Podtel is the perfect jumping-off point for exploring the desert's gullies, ravines and Neolithic dolmens by hiking, cycling or 4x4.

This state-of-the-art pod hotel is the first of many that the Spanish firm DistrictHive plans to open in remote locations round the world. The company's bold ambition is to revolutionise the luxury hotel experience and, indeed, this is not your everyday hotel. The eco-friendly and technology-drived podtel generates its own water from the air and its own electricity from solar energy, and has enough energy to power itself for four days without sun (should the sun stop shining in this sun-baked part of Spain), thus optimising resources and ensuring a reduction in the carbon footprint. The podtel is also transportable and, if moved, will leave no trace on the landscape.

DistrictHive describes the self-sufficient Colony as a 'human recharging station' and with nothing but mountains, sky and desert sand and rock for as far as the eye can see, it's certainly an appealing place to spend a few days disconnecting from the stresses of life and connecting with the beauty of nature. That said, the experience is more blissful solitude than strict asceticism. The compact-sized podtel is minimally decorated and luxuriously equipped, with room for up to four people, and divided into a bathroom, a kitchen-living-dining room and a bedroom. The Colony is controlled by artificial intelligence and does everything through an app on your phone, from opening doors to setting the lighting, temperature and aroma. You can even check in and out through the app and buy thoughtfully curated food and drinks from dispensers – no need to disrupt your desert solitude for anything or anyone. Outside, you'll find stargazing decks on hammocks to experience the beauty of the Gorafe Desert's night skies, some of the clearest in Spain.

The pod hotel promises total immersion in the Wild-West-esque Gorafe Desert.

The state-of-the-art capsule is 100-per-cent self sufficient. | Spain | **Europe**

NORTH AND
SOUTH AMERICA

DESERT GOLD

*Splendid isolation
in a modern desert oasis*

Death Valley

Beatty, Nevada

United States

airbnb.com

It all started with *Zabriskie Point*. Michelangelo Antonioni's 1970 movie about countercultural America sparked such a desert obsession in the Italian journalist Fabrizio Rondolino that while on holiday in Death Valley, where parts of the movie were filmed and one of the harshest places on earth, he bought land in what might be described as the middle of nowhere – or by Scotty's Junction, off Highway 267, to be more precise. Rondolino then commissioned the architect Peter Strzebniok, whose San Francisco firm, nottoscale, had created an award-winning prefabricated house, to build his dream home.

The result was Desert Gold, a partially prefabricated and artfully designed holiday home that sits on 80 acres (40 hectares) in an isolated patch of Nevada Desert, right behind Death Valley. The property offers unobstructed views to the desert and the mountains and sits on a plinth that elevates the building so that it appears to float above the desert. Its gold-hued exterior, which glows and reflects the ever-changing desert light throughout the day, complements its rugged sand and sagebrush-covered surroundings. Floor-to-ceiling windows invite the desert landscape, and plenty of natural light, inside and the airy and bright interiors feature minimal decoration with occasional pleasing pops of bright colour. A large living/dining/kitchen area extends onto a spacious deck that can be isolated from the rest of the house by pulling the glass doors shut.

It's a true desert hideaway; there's not another building in sight. The property's closest neighbour is the Shady Lady, a brothel turned bed and breakfast on Highway 95, seven miles away. A hermit, following a calling from above, used to live in a trailer on the next lot over – the chapel he built is still there, but his trailer has moved on.

The three-bedroom property sleeps up to six guests and is bookable through AirBnB. Splendid isolation is the main draw here, so expect to make your own entertainment, whether it's stargazing (there's no light pollution), watching the sun rise over the mountains from the hot tub or deck, or thumbing through books from the thoughtfully curated library. If you need to stretch your legs, you can go hiking round the surrounding desert or drive further afield to such Death Valley crowd pleasers as Zabriskie Point and Titus Canyon. Most of all, though, staying at this unique holiday home is all about unplugging from the outside world and spending quality time in a desolate, beautiful place.

North America | Nevada | This stunning home complements the stark beauty of the surrounding landscape.

HOTEL LUNA MYSTICA

Whimsical nostalgia meets stylish aesthetics in an otherworldly setting

New Mexico

25 ABC Mesa Rd, El Prado

United States | +1 575 613 1411

hotellunamystica.com

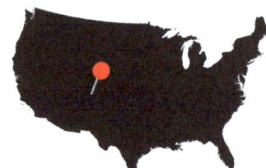

Surrounded by the most powerful landscapes in New Mexico, it's easy to fall under Taos' spell. From the immense Rio Grande Gorge to the west to the jagged peaks of the Sangre De Cristo Mountains that loom over the town, the high-mountain desert settlement of Taos alone earns New Mexico its apt moniker 'the Land of Enchantment'.

Fifteen minutes north of the heart of Taos, down a dirt road that gives way to a wide-open mesa, Hotel Luna Mystica casts its very own spell. Pull up to the sprawling grounds, where 21 vintage trailers glint and shimmer in the New Mexico sun, and you'll feel time wrap around itself. From Frida and Diego to Thelma and Louise, each trailer at Hotel Luna Mystica has its own personality and retains its vintage soul – all trailers were made between 1950 and 1972, the heyday of the great American road trip, if not the nation itself. You get the sense that, if these trailer walls could talk, they'd have quite a few colourful tales to tell.

This is a desert escape for the more adventurous traveller but you'll not lack for modern comforts. Each stylishly renovated trailer is equipped with its own deck, bathroom, shower, comfortable bed, kitchen facilities, and the all-important air conditioning unit to help take the swelter out of the desert heat (and heaters for the cold desert nights). For the more back-to-basics adventurer, there are tent sites, as well as primitive RV spaces without hook-ups.

Spend your days taking in the ever-changing views of northern New Mexico's stark desert plains and snow-capped mountains from the deck then, when night falls, light up the fire pit and watch the stars sparkle in the wide-open sky. An unconventional and fun-loving place like this lends itself to conviviality so you are likely to find yourself swapping stories with your fellow guests well into the wee small hours.

Hotel Luna Mystica feels a million miles from anything but you'll not want for liquid refreshment: Taos Mesa Brewery's Mothership is right across the road and renowned for its craft brews, as well as its live music events. There are also a number of great hikes nearby and a short drive will bring you to Manby and Black Rock hot springs where you can soak away any lingering stress, as well as the Taos Ski Valley, home to some of the country's best skiing.

Enjoy uninterrupted mesa views from your vintage camper.

Cosy campers come with bags of personality. | New Mexico **| North America**

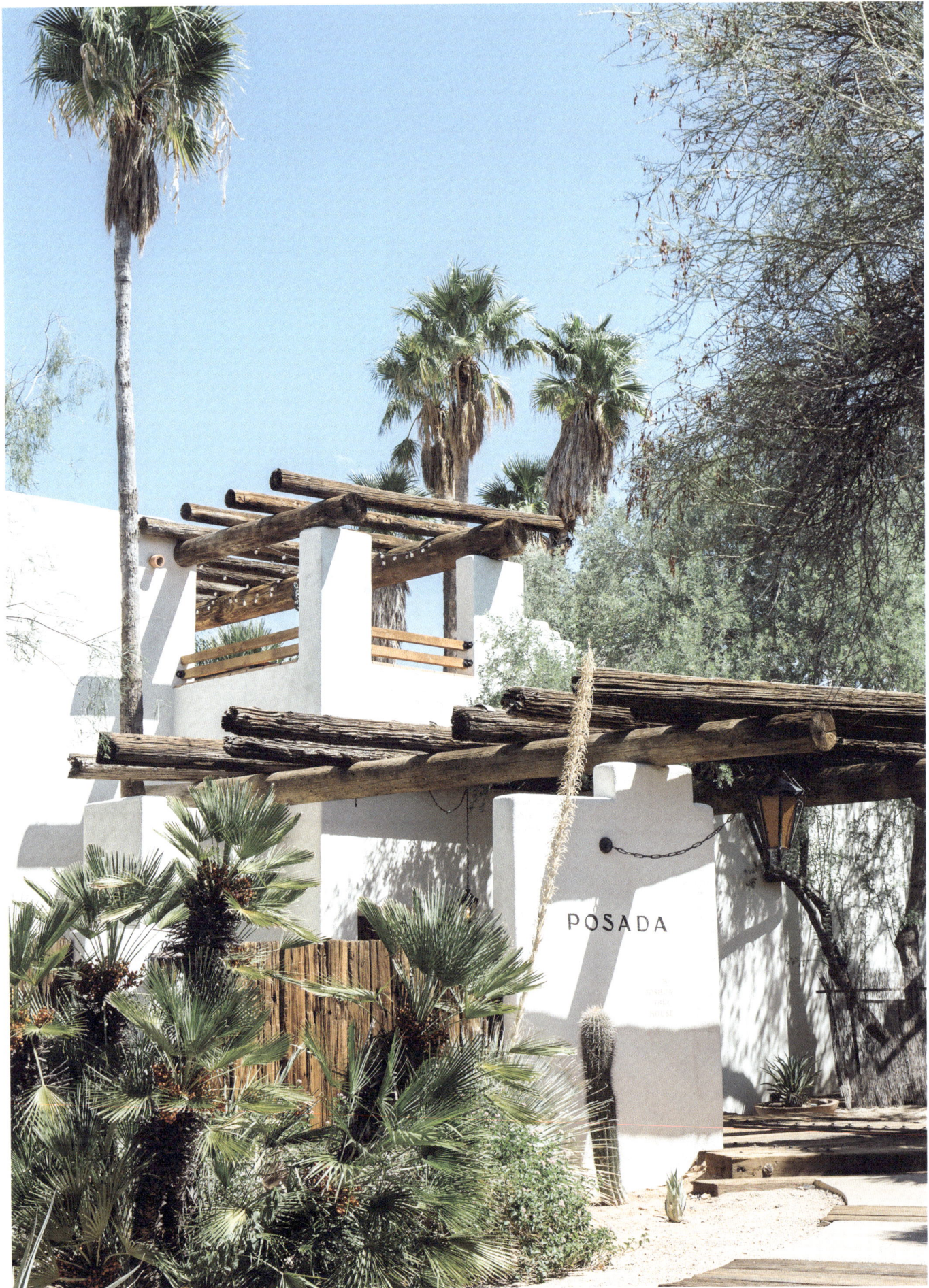

JTH TUCSON

Community and place are at the heart of this Arizonan desert retreat

Sonoran Desert

12051 W Fort Lowell Rd,
Tucson, Arizona

United States

thejoshuatreehouse.com

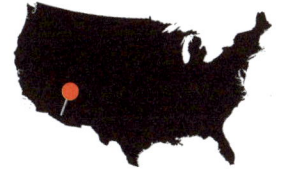

Bordering Saguaro National Park West in Tucson, Arizona, JTH Tucson is so much more than just a place to sleep. Covering 38 acres (15 hectares) of Sonoran Desert, and surrounded by saguaro and prickly pear cactus, this five-suite desert posada is a place to decompress while immersed in a magical landscape. It also offers access to artfully curated shared areas including a rock canyon pool, rooftop lounge, yoga and meditation room, dining patio, fire pit, and living room with a projector for shared movie nights. It's a space tailored to community, ideal for private gatherings, and you can book the entire inn to sleep up to 14 of your closest friends. The property works with a local chef who can be booked to create private meals in the well-equipped kitchen and it also has a list of people you can call on to further enhance your stay, from yoga teachers to masseuses.

JTH Tucson is the creation of Sara and Rich Combs, founders of the Joshua Tree House (located in Joshua Tree, California). That two-bedroom hacienda, which opened in 2015, set the tone for desert-escape decor and design and soon grew into a lifestyle brand encompassing a book, an online store, and several properties. In 2018, realising a long-held dream to open up an inn next to a national park, the couple purchased this abandoned adobe-style Arizona property to create JTH Tucson.

On their travels, prior to becoming the innkeepers, designers, entrepreneurs and authors they are today, the Combses felt they had to sacrifice either design or nature in the accommodation they offered, so they set out to create spaces that celebrate both. Fully immersed in its surroundings, JTH Tucson achieves that goal beautifully. Light, earthy tones give the inn a warm aesthetic

that melds with the landscape. Panoramic desert views are at every turn, from the spacious rooftop terraces to private patios, and are framed by picture windows that are perfectly positioned to capture desert sunsets and the starry night sky. While renovating the property, the designers brought several original features back to life, including the Saguaro rib ceilings and beehive fireplaces found in most suites. The centrepiece of the property is the split-level Saguaro Suite, which has a spa-like bathroom with a sunken tub and penny tiling.

When not enjoying the inn's pool, on-site trails and cactus garden, guests can head into Saguaro National Park – a few minutes' drive will take you to several trailheads. After working up an appetite, take a scenic drive through Gates Pass to Tucson and discover why it's a UNESCO-designated City of Gastronomy.

The inn is placed in a quintessential desert landscape of saguaro and prickly pear cactus.

The canyon pool is a favourite summertime hideout.

Original details of the 1970s-built adobe property have been lovingly brought back to life. | Arizona | **North America**

CAMP SARIKA BY AMANGIRI

Combine the solitude of the Utah Desert with the thrill of outdoor adventure

Canyon Point

1 Kayenta Road, Utah

United States | +1 435 675 3999

aman.com

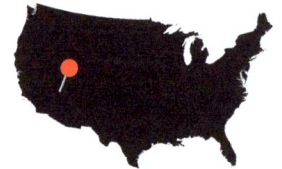

Amid an otherworldly, 600-acre (24-hectare) Wild-West landscape of rust-coloured sands, slot canyons and mighty mesas, Camp Sarika is the rugged, and even more secluded, offshoot of the ultra-luxe modernist icon that is Amangiri, one of the flagship properties of the exclusive Aman Resorts brand. While this tented retreat is just a short hike, or shorter drive, across the Colorado Plateau from Amangiri, Camp Sarika carves its own niche out of its dramatic surroundings, promising an intimate experience in the heart of the Utah Desert.

A 'camp' in the loosest sense of the word, this luxury retreat is comprised of a cluster of ten one- and two-bedroom canvas-topped pavilions, each with its own expansive terrace with private heated plunge pool and each placed gently in the landscape so far enough apart that you need never see another guest. Designed for the harsh environment, these tented pavilions can withstand strong desert winds and even the hottest days of summer and bone-chilling winters, making Camp Sarika the first year-round camp of its kind in North America. Each pavilion can be booked individually, or groups can book the entire camp for the ultimate in exclusivity. There's a shared lounge, restaurant and pool area and guests are welcome to make use of all the facilities, amenities and activities on offer at Amangiri.

Begin your day with an al-fresco yoga or meditation session and campfire breakfast. Afterwards, you can head out on an adrenaline-fuelled journey along the *via ferrata* – a climbing route with steel cables and ladders fixed into the rock face – explore the surrounding region by horseback, hike the gorges or float over soaring canyons by hot-air balloon. The natural wonders of the Grand Canyon, Bryce Canyon and Zion national parks, Grand Staircase-Escalante National Monument and Monument Valley Navajo Tribal Park are all within easy reach for a day of adventure, while Lake Powell and the Colorado River can be explored by boat – staff will personalise a tour to any part of Utah's landscape you want to experience. Back at the camp, after a dinner of southwest-inspired cuisine under the starry sky, swap stories while toasting s'mores on the fire pit.

The camp is cradled by steep buttes and dramatic mesas.

Camp Sarika's pool is an inviting oasis.

Each pavilion room opens directly onto the desert landscape. | Utah | **North America**

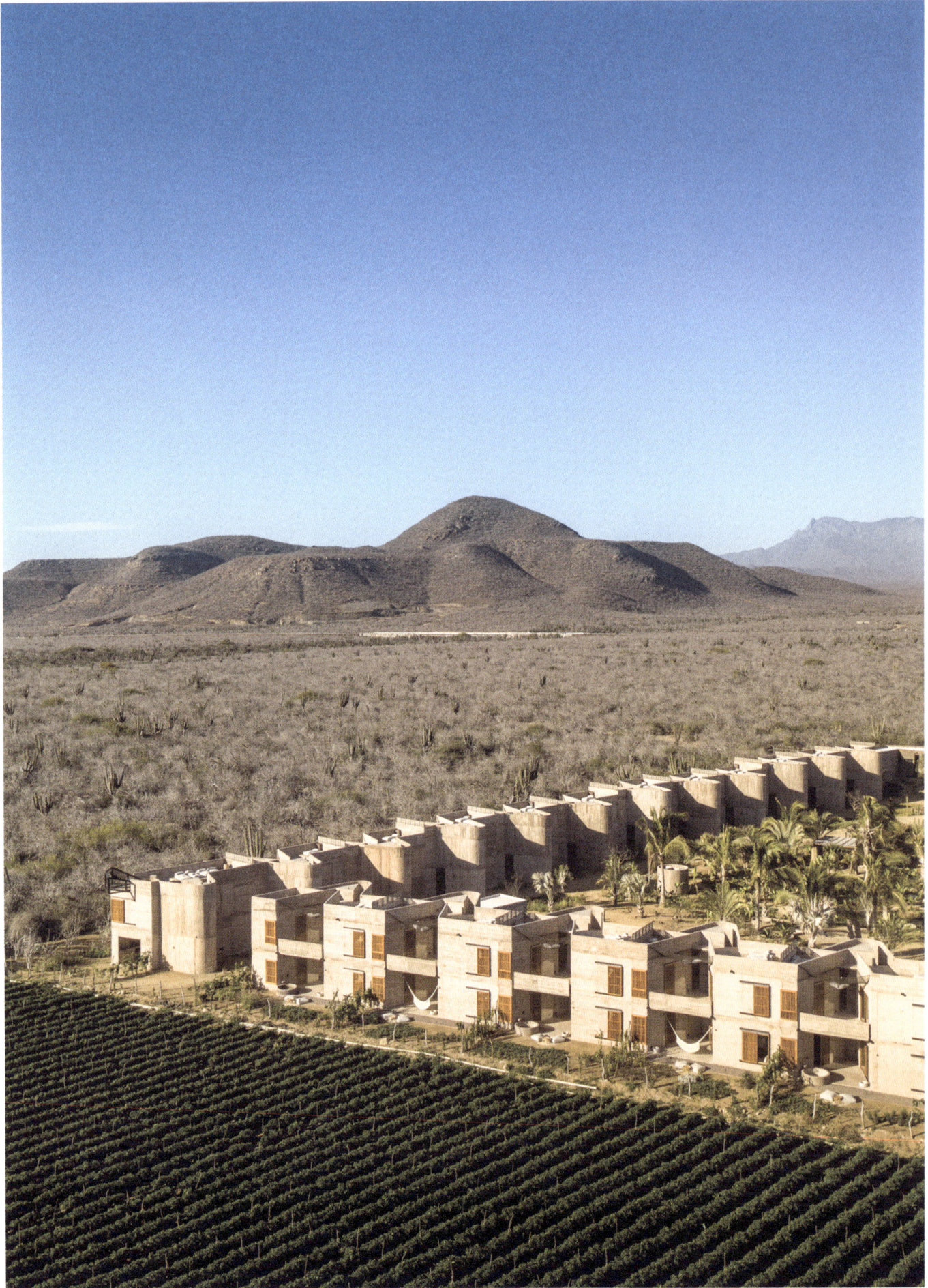

PARADERO TODOS SANTOS

A hotel rooted in community and adventure carves its own Baja niche

Baja California Sur

La Mesa KM 59 +3100, Carretera Todos Santos – Cabo San Lucas El Pescadero

| Mexico | +52 1 612 229 1764 |

paraderohotels.com

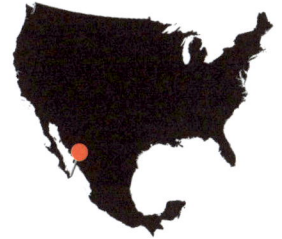

To get to Paradero Todos Santos, you must first turn away from the glitzy resorts and hedonistic temptations of Cabo San Lucas, then head north on a journey that ends in a farming community down a dusty country road. After about an hour, you'll arrive at a series of low-slung, undulating brutalist-inspired structures whose sun-baked desert tones blend effortlessly with a sandy landscape scattered with cactus and Mojave yucca. This is not your spring-breakers *baja*.

Paradero Todos Santos is an all-suite, experience-focused boutique hotel situated at the intersection of the desert, the ocean and the mountains. Existing in complete harmony with its surroundings, the property fancies itself less a hotel than a high-design landscaping project with luxurious suites.

The 35 suites are dotted along the perimeter of the property, promising unobstructed views from each. Inside, the design is minimalist, allowing the environment to take centre stage and blurring the boundary between indoor and outdoor. White cedar-wood furniture and neutral-coloured textiles made by artisans in Oaxaca and Guadalajara feel well matched to the landscape. Ground-floor suites have private plunge pools while those upstairs have roof terraces and suspended star net hammocks for elevated views of the cactus forest, farmland and mountains. At the edge of the property is a half-moon-shaped infinity pool with swim-up bar; a below-ground spa rooted in traditional Mexican healing practices including sound healing and Temazcal ceremonies; and an open-air restaurant. It has a

traditional Oaxacan clay tortilla oven and makes good use of ingredients sourced from the property's own 100,000 square-foot (9,000 square-metre) botanical garden, as well as the best of the region's organic and sustainable ingredients.

Making authentic connections with community and land lies at the heart of the Paradero experience and local guides are on hand to help guests become immersed in nature while they share their generation's worth of knowledge of and reverence for the land. You're encouraged to get off the property and into the landscape and a myriad guided experiences is on offer to help you do so, from surfing at nearby Playa Cerritos, mountain biking along the Pacific coast and hiking over sand dunes, to a hands-on sustainable farming class.

Designed by architects Ruben Valdez and Yashar Yektajo, the hotel's beige concrete exterior blends with the surrounding terrain.

Todos Santos is a designated Pueblo Magico, or magical town, for its natural beauty and cultural heritage.

Guests can explore the desert and the ocean through activities from hiking to surfing. | Mexico | **South America**

PRISTINE CAMPS

Sustainability is at the heart of this luxury glampsite on the shimming Argentinian salt flats

Salinas Grandes

Salinas Grandes, Jujuy	
Argentina	+54 9 11 6033 7460
pristinecamps.com	

Pack your shades for your trip to the vast, blindingly white salt flat of Salinas Grandes, in northwest Argentina's Jujuy province. This dazzling landscape – a lake that dried up long ago, leaving behind a huge, thick crust of salt – is one of the country's most spectacular sights, particularly on a clear day when a bright blue sky contrasts with the stark, silvery white cracked and crusted ground.

Visitors can enjoy the silence and clear skies of Salinas Grandes by spending the night in an igloo-inspired dome on the salt flats, courtesy of Pristine Camps, the first in what will be a boutique chain of luxury camps around the country. In order to maintain its sense of seclusion and exclusivity, the camp

has just four domes, named Sun, Moon, Star and Earth, and a capacity of no more than ten guests. You can choose between an Executive Suite or a larger Premium Suite, which has an outdoor deck, perfect for stargazing sessions. Far away from the light pollution and trappings of the modern world, the night skies couldn't be clearer.

While staying here can feel like a step back to a simpler time, Pristine Camps has its eye on the future. Sustainability is a core pillar and each of the four domes run on 100-per-cent renewable energy and are raised off the crusted earth on wooden platforms to prevent erosion and preserve the natural environment as much as possible. Social sustainability is another pillar

and so the camp contributes to the local economy and culture by training and employing young people from a local co-operative to introduce guests to their traditions and artistry. Local producers are given priority whenever possible: menus feature local ingredients, paired with wines sourced from the nearby Colomé vineyard.

Daily excursions are included in rates and offer an even more immersive experience of the region's culture and environment. Activities include archaeological excursions, visits to small villages, and guided astronomy sessions. Open all year round, Pristine Camps is at its best from November to March when summer rains replenish the salt and create a mirror effect reflecting the brilliant blue sky on the surface of the flats.

The salt flats of Salinas Grandes are dazzling. | Argentina | **South America**

FOLLY JOSHUA TREE

A modern off-grid hideout in the heart of California's Mojave Desert

Mojave Desert

Twentynine Palms, California

United States | +1 954 871 9631

follycollection.com

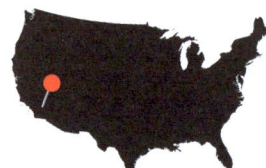

Named for its twisted, spiky trees whose limbs stretch out as if in supplication, Joshua Tree National Park is characterised by rugged rock formations and stark Mojave Desert landscapes. Artists, writers and musicians have long been inspired by the park's ethereal beauty and supposed spiritual energy. More recently, desert lovers have been renovating cabins and abandoned buildings to share the area's magic with travellers and new transplants.

Just outside the park, Folly Joshua Tree is one of the best. Two cabins stand in the stark desert like a pair of monumental sculptures, their rusted metal exteriors merging with the earthy hues of the desert. Its open-air spaces blend indoor and outdoor living with an emphasis on innovative

minimalist design. Folly invites guests to disconnect from everyday life, immerse themselves in off-grid living, and connect with their surroundings.

Run on solar power, Folly Joshua Tree fits up to six people, but is ideal for two to four. The modern and refined main cabin (a renovated and revived 1950s homestead cabin) blends natural stone, warm plywood accents and dark furnishings, and has a living and dining room, small kitchen, lofted bedroom and shower with a picture window. In the smaller second cabin (a new build) guests can climb the ladder to find the stargazing suite, where a queen bed laid out in the ceilingless loft offers the chance to sleep under the stars, or soak up the sun during the day – a cover can be drawn over the space in the rare

event of rain. There's also a projector for watching movies on the back wall. Between the two cabins, an expansive outdoor area with soaking tub, fire pit, bocce ball court, daybeds and hammocks invites guests to relax while admiring the landscape and cotton candy sunsets.

Folly Joshua Tree is a place to disconnect and decompress, but when adventure calls, the trails of Joshua Tree National Park and the cafes, bars, markets and quirky attractions of the offbeat and artsy town of Joshua Tree are all just minutes away. Visit in the spring and you'll see the colourful spectacle of wildflowers blooming in the park, from hot-pink beavertail cactus to neon-yellow grizzlybear pricklypear cactus and vermillion desert paintbrush.

Switch off the outside world and immerse yourself in one of the world's most striking landscapes.

This tiny home is solar powered and utilises space precisely. | California | **North America**

HAWK & MESA

A 'cowboy modern' southern California desert retreat

Mojave Desert

Yucca Valley, California

United States

hawkandmesa.com

In California's Mojave Desert, the Hawk & Mesa house is tucked into a canyon on the outskirts of historic Pioneertown. Built to look like a typical 1800s Old West town, Pioneertown was an elaborate living movie set where more than 50 Western movies starring the likes of John Wayne and Gene Autry were shot during the 1940s and 1950s. After the film studios stopped using Pioneertown as a back lot, it was transformed into a real town and is today filled with boutique stores, music venues, bars and restaurants.

Pioneertown was an important source of inspiration for Hawk & Mesa, where the design takes its cues from the metal pole structures used by local ranchers to shade horses and the weathered wood homestead cabins of the area.

Architect Jeremy Levine describes the aesthetic of reclaimed lumber and exposed steel as 'cowboy modern'.

Set on 120 acres (50 hectares) of boulders, ancient desert oaks and spiky Joshua trees, the two-bedroom house feels as though it belongs entirely to its surroundings. The slope of the roof mirrors the slope of the surrounding hills and the exterior siding is made of reclaimed lumber that has the weathered look of the old homestead cabins and blends into the desert's colour scheme, while floor-to-ceiling windows invite desert views inside. Every room in the house has large sliding glass doors that open up to a shaded outdoor terrace that echoes the typical shaded porches of Pioneertown. Sustainability was a driving force behind every design decision, so the location was carefully

chosen to minimise impact, produce little waste and take advantage of the sunlight and canyon breeze in order to light and cool the house.

With no immediate neighbours, guests enjoy a profound sense of solitude and serenity, and plenty of room to breathe. From rock climbing on the sprawling property's grounds to hiking in Joshua Tree National Park (a 25-minute drive) and Big Morongo Canyon in the Sand to Snow National Monument (30 minutes), outdoor activities here are truly world class. After a day exploring, return to sundowners by the fire pit or a soak in the hot tub or cool off in the cowboy tub, a repurposed galvanised stock tank. Watch for hawks, take in views of the mesas from your comfy hammock and wait for stars to light up the night sky.

This sustainability-minded retreat has big views and a small footprint. | California | **North America**

THE PERCH

Retro meets modern at this fully off-grid `cabin-inspired´ home

Chihuahuan Desert

1400 Reed Plateau, Terlingua, Texas

United States

perchterlingua.com

Secreted off a bumpy dirt road, ten minutes from Terlingua Ghost Town, this three-bedroom house is set on the cliff edge of a 60-million-year-old limestone plateau. The sweeping view is the star attraction here, and while there is no TV to distract you, you may get lucky and spot a few of the desert creatures that make their way through The Perch grounds - such as ringtail, desert foxes, roadrunners, coyotes and aoudad (Barbary sheep).

Nestled on 20 acres (8 hectares) of pristine Chihuahuan Desert, The Perch overlooks historic Terlingua and into the expanse of the Chisos mountain range of Big Bend National Park. It's the perfect spot for couples, small groups or anyone who wants to enjoy the seclusion and beauty of the Big Bend region.

Visitors can pass their time exploring the old mining history of Terlingua's historic district, browse the shops and art galleries, and enjoy a taste of Texas-meets-Mexico at one of the restaurants – *huevos rancheros* for breakfast is a must. The town is sandwiched between Big Bend National Park and Big Bend Ranch State Park, which together encompass more than a million acres (over 400,000 hectares) of protected outdoor recreation land for hiking, taking river trips on the Rio Grande and exploring the Chisos Mountains. Any time is great to visit but the spring (March/April) and autumn (November/December) are most appealing since the weather is comfortably cooler – spring has the added bonus of native desert plants, including yellow desert marigold and Big Bend bluebonnets, blooming throughout the region.

Back inside the house, design enthusiasts will love the architecture, which mixes modular shipping containers, traditional southwestern stucco and modern rusting weathering steel in a unique style that blends the comfortable airiness of midcentury Japanese summer homes with the edginess of brutalism and retro sci-fi. Soft tones and clean lines evoke a little Scandinavian pre-fab for good measure. When darkness falls, head out onto the back porch and enjoy the twinkling stars as they put on a show overhead. The Perch is a certified Dark Sky Friendly property, which meets standards set by the McDonald Observatory's Dark Sky Initiative in partnership with the International Dark Sky Association. This spirit of respect for the environment is in keeping with the owner's ethos of minimising the property's footprint on its beautiful landscape.

This dark sky-friendly property is committed to preserving Big Bend's magical night sky.

Sitting high on a plateau, the home is precisely designed to maximise space and views. | Texas | **North America**

NAYARA ALTO ATACAMA

In the heart of the Atacama, this adobe wilderness lodge embraces its surroundings

Atacama Desert

Camino Pukará, Suchor s/n,
San Pedro de Atacama, Antofagasta

| Chile | +56 (2) 2912 3900 |

nayaraaltoatacama.com

Blending effortlessly into its terracotta-coloured surroundings, the low-slung, adobe-walled Nayara Alto Atacama is an oasis at the heart of the driest place on earth, the Atacama Desert. An inviting and secluded hideaway in an unforgiving environment, it's an ideal basecamp for those with a sense of adventure.

Nayara Alto Atacama is tucked into the Catarpe Valley in the Salt mountain range, near the pre-Columbian ruins of Pukará de Quitor and the San Pedro River. While it's just minutes from the travellers' basecamp of San Pedro de Atacama, it may feel like a world away. The 42-room lodge's design takes its cues from the region's traditional settlements and the building patterns of the Likanantaí people – and it spares little detail.

The sprawling grounds feature the traditional landscaping of the altiplano and include the Andean Garden where flourishing native crops are a rebuke to the challenging terrain – as you stroll round, be sure to greet the lodge's herd of friendly llamas in their nearby corral. Indoors, neutral colours and crafts handmade by local artisans add warmth and character; each spacious room has its own private terrace with views of the interior gardens, the Catarpe Valley or the Salt mountain range.

Vegetables from the Andean Garden make their way onto the menu at Ckelar Restaurant, which is influenced by traditional Atacama and Chilean cuisine. Sip on the restaurant's strong selection of Chilean wines while sampling such local dishes as

guanaco meat; fresh quinoa salads; corn cake with goat cheese and chañar berries; and *charquicán*, an Andean style of stew.

The lodge's location provides easy access for adventurous excursions to the desert's high altitude lagoons, otherworldly salt flats, geothermal fields and moon-like valleys, while its outdoor and indoor spaces lend themselves to total relaxation – hop between the six pools and restore body and mind with a treatment in the Puri Spa. However you spend your days, when evening falls, you'll want to experience the magic of the Atacama's clear night skies at the lodge's own astronomical observation deck before winding down by the bonfire, pisco sour in your hand, as stars sparkle overhead.

This adobe-walled lodge connects guests to their environment through cultural immersion and adrenaline-pumping adventures.

The earthy tones of the lodge's cool interiors provide a calming oasis in the world's driest desert. | Chile | **South America**

TIERRA ATACAMA

*Dazzling architectural delight
near San Pedro de Atacama*

Atacama Desert

Calle Séquitor, San Pedro de Atacama,
Antofagasta

| Chile | +56 800 914 249 |

tierrahotels.com/atacama

The stars seem to shine more brightly at Tierra Atacama, which has been carefully placed in the Atacama Desert for sublime stargazing sessions, whether your preferred style of astronomy is with a giant telescope or a giant glass of wine in your hand.

On the edge of the small town of San Pedro de Atacama, and set on a rugged, arid plateau in northeastern Chile's Andes, Tierra Atacama is a perfect hideaway for the intrepid explorer. The list of possible adventures here runs the gamut from visiting the Tatio geysers at sunrise and hiking through the desert under a full moon to climbing the Lascar Volcano and biking to the Quebrada del Diablo (Devil's Canyon). Pair your adventures with rejuvenating treatments at the Uma Spa where you can soak in the waters of the al fresco

hot tub and infinity pool while soaking up views of the Licancabur Volcano. Surrounded by flamingo-dotted salt flats, otherworldly valleys, natural hot springs, and six volcanoes, it would be easy for a hotel to be overshadowed by its spectacular scenery, but Tierra Atacama shines as brightly as that clear, star-studded desert sky. Originally a cattle corral, where drovers paused to rest after crossing the Andes, the property was transformed by the Chilean landscape architect Teresa Moller, who preserved the ancient carob and chañar trees and restored the adobe walls. The single-level, desert-modernist-style building houses 32 rooms, which feature white walls, four-poster beds with roll-down wooden shutters, and floor-to-ceiling windows that perfectly frame the desert scrub and Andean peaks. Locally crafted furnishings add to the authentic Atacama ambience

while cowhide rugs and alpaca throws provide a touch of cosiness against the cool desert nights. Private terraces and outdoor showers offer yet more perspectives on the spectacular surroundings.

Local produce is the star at Tierra Atacama's restaurant, some of which is sourced from the on-site garden, just steps away from your table. Breakfast is a generous buffet with plenty of fresh-baked bread and coca tea to set you up for a day of adventure, while daily changing, three-course set lunch and dinner menus champion north Chilean beef and lamb. This is Chile, so naturally, the wine list is impressive – you can choose from some of the country's best sauvignon blancs, merlots, pinot noirs and syrahs. And the pisco sour here is the best you'll ever sip in the desert.

Relax and rejuvenate in the shadow of the Licancabur Volcano. | Chile | **South America**

EL COSMICO

This Marfa retreat is a modern-day campground that radiates utter coolness

Chihuahuan Desert

802 S Highland Ave, Marfa, Texas

United States | +1 432 729 1950

elcosmico.com

One of the nation's coolest desert escapes, the 21-acre (8.5 hectares) El Cosmico boasts camera-ready aesthetics with its restored and revived vintage trailers, stylish yurts, safari tents and 56-foot (17-metre) tepees. At El Cosmico, travellers are encouraged to ditch their devices, unplug from the modern world and spend their time biking, stargazing, attending creative workshops or exploring the small-town charms of Marfa (within walking distance), where the art scene draws visitors from around the world. Donald Judd founded the minimalist Chinati Foundation art museum in Marfa in 1986, which catapulted the town into the national spotlight – if Elmgreen and Dragset's *Prada Marfa* art installation hasn't yet crossed your feed, are you really on Instagram?

Bring your sense of adventure to this hip outpost tucked into the vast expanse of west Texas. Hotelier Liz Lambert (known for Austin's stylish Hotel San Jose and luxe Saint Cecilia) envisioned El Cosmico as a communal oasis – think post-hippy commune. She collaborated with architecture firm Lake/Flato and a creative team of artists and designers to give each midcentury trailer a new life with its own distinctive style and decor, from bold colours and 1960s aesthetics to mimimalist white interiors and wood details. The artistic, grown-up camp is an elevated version of 'van life', but be prepared to expand your comfort zone: some trailers, as well as the yurts, safari tents and tepees, share facilities in the central bath house. Spacious Bushtec Tents, on the other hand, blend cabin life with stylish comfortable amenities, coming with vintage soaking tubs, spacious

front and back decks, heat/AC and electricity, as well as a private outdoor shower. At either end of the scale there's the roomy, well-appointed Cosmica Kasita microhome and the campground where for a nominal fee you can pitch a tent, making the El Cosmico experience available to pretty much anyone willing to make the trek there.

If camp life is not for you, opt for a stay in the 1930s Brite Building, El Cosmico's annex in the heart of downtown Marfa. The 3,800 square-foot (353 square-metre) flat blends art deco and West Texas minimalist design and is located above the Ayn Foundation (which houses several works by Andy Warhol) and has two bedrooms with king beds and private bathrooms, as well as generous living, kitchen and dining rooms.

Tepees and trailers promise a galactic stay in Marfa.

Thoughtfully restored vintage trailers provide the perfect retreat for the nomadic spirit. | Texas | **North America**

CASA SILENCIO

Immerse yourself in the soul of mezcal at this distillery hotel

Oaxaca Valley

Xaagá, Oaxaca

Mexico | +52 951 190 40 77

casasilencio.com

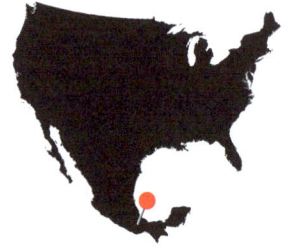

Spirits enthusiasts will love Casa Silencio, which is situated at the home of the El Silencio mezcal distillery and dedicated to the appreciation of mezcal, that powerful, slightly sweet spirit that represents Oaxaca in a glass.

Casa Silencio is located in the windswept Valley of Silence, at the end of an agave-lined dirt road out of the town of Xaagá – a small pueblo on the outskirts of the Mitla archaeological site, which has influences from Mexico's Zapotec and Mixtec communities. The six-room boutique property is an off-the-beaten-track, near-ceremonial space for mezcal enthusiasts who want to fully immerse themselves in the culture of the spirit and taste rare editions of El Silencio, the hand-crafted mezcal brand founded by Fausto Zapata and Vicente Cisneros. Designed by the Mexican architect Alejandro D'Acosta, the sustainability-focused property's architecture draws on indigenous buildings traditions of the area by using rustic, subdued materials such as reclaimed wood and stone, terracotta tiles and compressed rose-hued clay to help the hotel blend into its natural surroundings. The interiors, however, feel sleek and contemporary. Each of the hotel's two-floor guest rooms is decorated with distressed leather detailing, velvet upholstery and locally crafted accents such as copper lamps and wool rugs hand woven in a nearby village. Living rooms have wood-burning stoves, while bathrooms boast enormous black-tiled walk-in showers, monolithic stone sinks and, in one case, walls crafted entirely from recycled El Silencio bottles. The heart of the property is the open-air dining room, where contemporary takes on classic Oaxacan dishes – such as rib eye with black *chichilo* sauce and fish tacos with *macha* sauce – are served at a long communal basalt table with sweeping views of the surrounding mountains, agave-studded valleys and star-strewn desert sky. The dining room is just a short way from the solar-powered *tahona*, a half-tonne stone wheel that crushes the roasted agave hearts before the pulp is placed in giant wooden barrels to allow the sugars to turn to alcohol. Each day, guests have the opportunity to lend a hand in the mezcal production, or can simply take it easy and enjoy it straight from the bottle, round the fire pit.

Find this eco-minded retreat at the home of Mezcal El Silencio.

Artisans from the nearby village of Xaagá created the majority of the hotel's decorative elements. | Mexico | **South America**

MODERN TAOS HOUSE

Off-grid, sustainability-focused luxury in Taos

New Mexico

El Prado, New Mexico

United States | +1 505 470 8125

moderntaoshouse.com

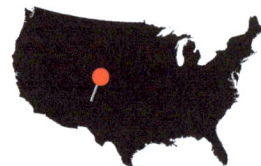

The show-stopping sunrises over the Sangre de Cristo Mountains deserve a retreat to match. That's where the Modern Taos House comes in. This two-bedroom, two-bathroom, solar-powered and rainwater-fed house sits close to the Rio Grande Gorge near Taos, New Mexico, and offers panoramic mountain views and unmatched tranquillity.

With endless views of the Sangre de Cristo Mountains and sagebrush-dotted desert plain on offer from each of the house's expansive windows, you may find it difficult to leave the comfort of the home. But here, with the Rio Grande del Norte National Monument as your backyard, you'll want to do plenty of exploring. Encompassing 242,555 acres (98,158 hectares) of public land, the landscape of the national monument features rugged, wide-open plains dotted with volcanic cones and cut by steep canyons and rivers with surging rapids. From whitewater rafting and fishing to hiking and mountain biking, the area offers a wealth of outdoor activities. After a day of recreation outdoors, you can head back to town and explore the Taos Plaza and Downtown. Browse the unique New Mexican art galleries and boutique shops, then take in a live music show at the Taos Mesa Brew Mothership, which is connected to a brewery with about a dozen beers on tap and just a stone's throw from the Modern Taos House.

While the attractions of Taos are easily accessed, the house has plenty of open space round it, creating the sense of a secluded sanctuary that is ideal for those seeking a little rest and relaxation in a beautiful location. Indeed, everything you need is right there. The house is designed with an open layout, featuring stylish furnishings and high-tech modern appliances such as smart TVs and Bluetooth sound systems. The sleek industrial-style kitchen offers plenty of space and equipment for preparing feasts and has a huge hand-polished concrete island and six bright red bar stools for gathering – the custom-built dining room table even doubles as a ping-pong table. Comfy beds promise an excellent night's sleep, but, come morning, those spectacular sunrises over the Sangre de Cristo Mountains are bound to pull you up out of bed, ready for another day's exploration.

This house is perfectly placed for outdoor adventures from skiing to white-water rafting.

The home runs mostly off solar power and rainwater. | New Mexico | **North America**

ROCK REACH HOUSE

*The views rock
at this High Desert home*

Mojave Desert

55798 Acoma Terrace,
Yucca Valley, California

United States | +1 760 299 5010

homesteadmodern.com

Rock Reach House seems to float high above a rugged terrain characterised by a jumble of giant weathered boulders and ancient juniper, piñon and desert oak trees. This modern steel-and-glass house sits at just over 4,000 feet (120 metres) above sea level in the private community of Rock Reach within the Mojave Desert town of Yucca Valley.

The two-bedroom home invites its stunning surroundings inside through sliding glass doors in each room, each leading out to private patios, and a long deck that extends from the living area to a kitchen that is fully furnished with high-end appliances. Outside, there is a wood-burning fireplace, shower, yoga deck and platform for

sleeping under the starry desert sky. You can keep warm during the cool desert nights with a soak in the outdoor hot tub and cool off during the day with a dip in the repurposed galvanised stock tank 'cowboy tub'.

The home is so precisely tailored to meet a guest's every need and whim that it may seem surprising to learn that it was completed in just eight weeks in 2009. Conceived in collaboration with Palm Springs-based o2 architects, Rock Reach House was the first home built by prefabricated home developer Blue Sky Building Systems using the Blue Sky Frame, an innovative light-gauge steel frame system. It enhances sustainability by minimising transportation volume and

waste. Inspiration for the building's pure structure of columns and planes was found in Le Corbusier's 1914 Dom-Ino House.

Rock Reach House is part of the Homestead Modern collection of design-centric private holiday homes in the High Desert, each of which offers seclusion with the promise of outdoor adventure on your doorstep. The house sits less than a 30-minute drive from Joshua Tree National Park, so you can be exploring its surreal and whimsical landscapes soon after lacing up your hiking boots. You're also a stone's throw from the quirky charms and honky-tonks of the former Western movie set Pioneertown.

This modern steel house is perched amid a boulder-strewn High Desert landscape.

Sunbathe or stargaze from the deck or outdoor tub. | California | **North America**

UNDER CANVAS

Upscale camping in the quintessential desert setting of Moab

Colorado Plateau

13784 US-191, Moab, Utah

United States | +1 888 496 1148

undercanvas.com

Picture the archetypal desert landscape and it's likely that the country surrounding Moab comes to mind with its red dust, burnt cliffs and water-and time-worn slot canyons. This is Edward Abbey's 'centre of the world, God's navel, the red wasteland', as he wrote in *Desert Solitaire*. Allow that raw red desert to seep into your bones and a visit here can feel transcendent.

Right in the heart of this powerful southwestern landscape is Under Canvas Moab, an elevated take on camping and the perfect jumping-off point for both Arches National Park (a ten-minute drive away) and Canyonlands National Park (25 minutes). Spread over 40 acres (16 hectares) of deep canyons and towering plateaus, this property's

40 beige canvas-covered, safari-style tents create a minimal impact on their dramatic surroundings. Under Canvas operates several elevated outdoor camps across the United States, primarily near national parks, and all are intentionally designed to minimise disturbance, maximise open space, and flow with the natural topography of the land.

Each airy tent comes with West Elm furnishings, including plush king-size beds, and most have en suite bathrooms. A family-friendly desert escape, Under Canvas Moab offers a selection of tents with adjacent tepee-style tents where the kids can sleep. Keen astronomers should book the Stargazer Tent, which features a window above the bed for viewing the night sky. It has been

kept free of light pollution thanks to the camp's collaboration with the International Dark Sky Association.

With two of America's greatest national parks, and a much greater area of public lands, within reach, it's easy to answer the call of the outdoors. Guests can choose from a tailored list of guided experiences, which include hiking in Arches and Canyonlands, canyoneering in the Sand Flats Recreation Area, 4x4 safari tours, river rafting and mountain biking. After a day of adventure, you can cook up your dinner on your tent's wood-burning stove, or at one of the communal grills, before gathering round the campfire to toast marshmallows and swap stories with your fellow guests.

Under Canvas is the perfect perch from which to plan your Utah Canyon Country adventures.

Safari-style tents promise a close connection with the outdoors. | Utah | **North America**

AUSTRALIA

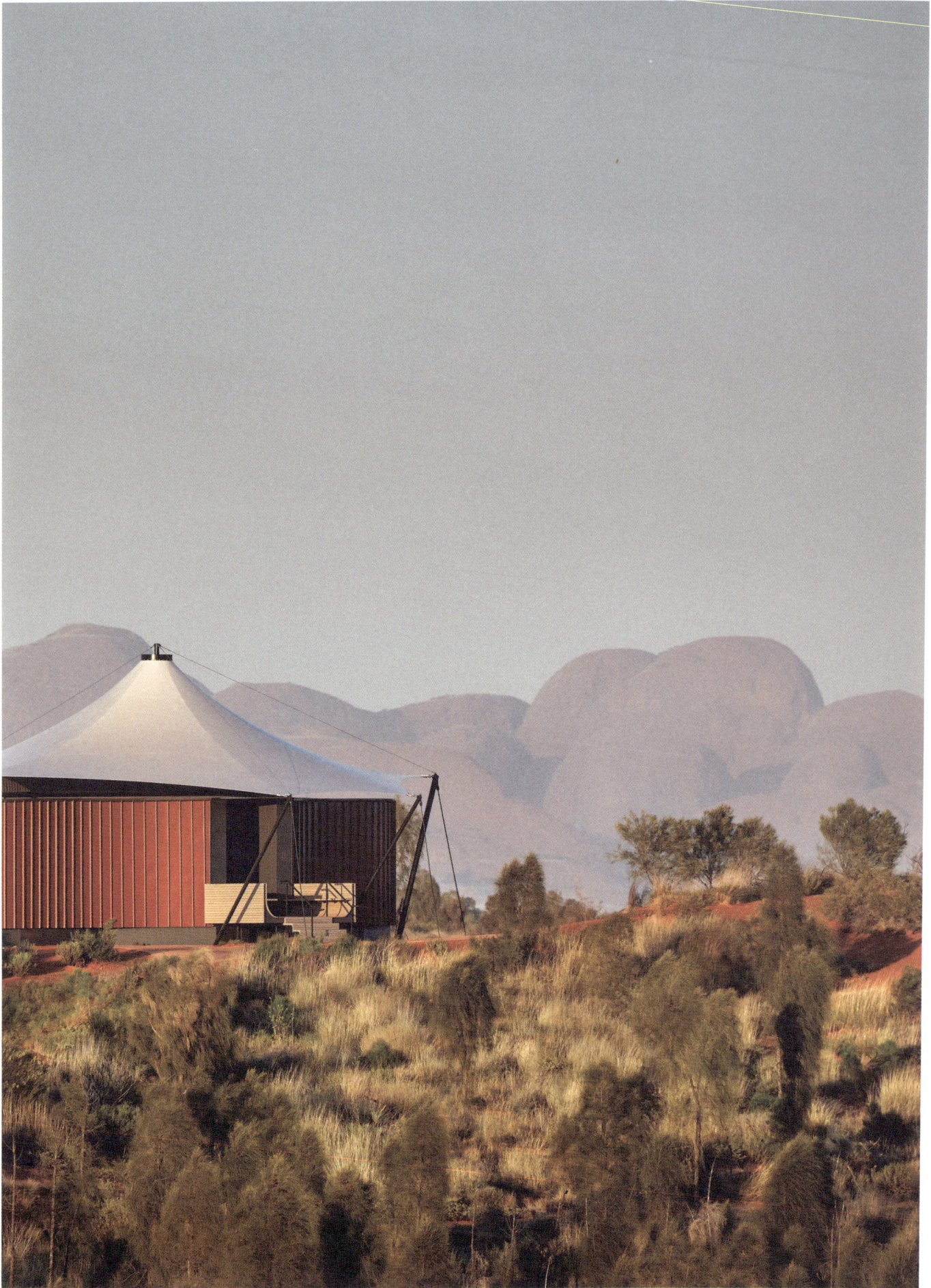

LONGITUDE 131°

*Spectacular views and glamorous
seclusion in Australia's spiritual heartland*

Simpson Desert

Yulara Drive, Yulara	
Australia	+61 2 9918 4355
longitude131.com.au	

Lying at the gateway to Uluru-Kata Tjuta National Park in Australia's Northern Territory, and surrounded by the rugged rust-coloured plains and dunes of the Simpson Desert, Longitude 131° commands unparalleled views of the iconic monolith Uluru – and you don't even have to leave your bed to enjoy them.

Guests at this romantic, secluded retreat in the spiritual and geographical heart of Australia stay in 16 luxury tents, which are safari-style pavilions topped with flowing white canopies that make a gentle impression on the landscape. Tents are raised above the sandy desert floor in order to minimise environmental impact and floor-to-ceiling windows offer panoramic views of Uluru and its ever-changing colours from the feet of the plush, organic linen-draped beds. You

can get even closer to nature by requesting a 'swag turndown' service and sleeping out on the roomy balcony in the cool night air under a blanket of stars. Each tent is named after a notable Australian pioneer or explorer and adorned with relevant memorabilia, such as letters or photos from their daring adventures, and custom-made furniture from Australian designers is complemented by colourful artworks by local indigenous artists. The two-bedroom Dune Pavilion offers the ultimate in outback luxury with separate living and sleeping areas, deep soaking tubs and a private plunge pool. It's also Australia's only accommodation that boasts views of both Uluru and Kata Tjuta.

At the heart of the resort is the Dune House, where you can sip sundowners at the convivial help-

yourself bar. A fusion of modern Australian and indigenous flavours is served for dinner either inside the Dune House or under the stars at Table 131°, which is perched atop a sand dune.

Soaking up the views of rock is the main attraction here, but the resort offers a wide selection of guided touring experiences, from an early morning walk round the base of Uluru or a hike to the Karingana lookout, to a camel ride across the dunes or traditional dot-painting lesson with Aboriginal artists. After a day spent exploring the Red Centre's vast outback landscape you can retreat to Spa Kinara for a rejuvenating treatment or simply enjoy the serenity and stillness of your surroundings while cooling off from the searing desert heat in the plunge pool outside the Dune House.

The closest hotel to Uluru offers unimpeded views of the sacred rock. | Simpson Desert | **Australia**

MT MULLIGAN LODGE

A rugged and intimate escape in the shadow of majestic Mount Mulligan

Queensland

499 Thornborough-Kinsgborough Road, Mount Mulligan

Australia	+61 74 777 7377

mountmulligan.com

Outback solitude and adventures await at Mt Mulligan Lodge, situated in the vast and rugged North Queensland backcountry, a 2.5-hour four-wheel drive or 35-minute helicopter ride from Cairns. Tucked away in a 69,000 acre (28,000 hectares) working cattle station, home to about 2,400 head of Brahman-cross cattle, this luxurious and intimate property sits below the spectacular 11 mile (18 kilometre) sandstone ridge of Mount Mulligan. It is to Aboriginal people as Ngarrabullgan and is a natural outback wonder that is ten times larger than Uluru, yet far less known.

The few travellers who make it to this out-of-the-way escape will find a captivating mix of cinematic landscapes, contemporary comfort and rich blend of indigenous and colonial history – inhabited by the Djungan people for almost 40,000 years, the landscape still bears traces of the late 19th-century gold rush and 20th-century coal rush. The remains of old Mount Mulligan township, now a ghost town, and mine can be explored while on the property.

The lodge's spacious suites and pavilions overlook a sparkling weir and have a total capacity of just 16 guests. All boast majestic views of the ever-changing colours of Mount Mulligan and have deep corrugated iron baths on their private decks. Those seeking a (relatively) rawer outback experience can book one of the property's outback tents on the other side of the weir. They are nestled amongst Australian eucalyptus woodlands- don't be deterred by the name, these 'tents' come with full-size beds, air conditioning and well-stocked minibars. No matter which you choose, all accommodation comes with its own electric buggy for impromptu explorations of the sprawling station.

Mt Mulligan is an all-inclusive lodge and as well as three gourmet meals a day, each with an emphasis on locally sourced seasonal produce, guests can choose a daily curated experience, which runs the gamut from adrenaline-pumping ATV adventures to gently floating downstream on a paddleboard, barramundi fishing in the weir, or witnessing the outback spectacle of cattle mustering. After a day of adventure, cool off in the infinity pool (or keep cosy by the main pavilion's fireplace), then head to the Sunset Bar to watch the outback sky turn shades of orange and red with a glass of outstanding Australian wine in your hand.

In the Queensland outback, this lodge sits in desert-like surroundings below the sandstone ridge of Mount Mulligan.

Guest suites offer a refined take on rural living. | Queensland | **Australia**

CREDITS

Cover:
Camp Sarika by Amangiri
© Courtesy of Aman

Introduction:
Wolwedans Collection
© Wolwedans Collection

AFRICA
© Alex Teuscher for Zannier
Hotels Sonop

p.10-15
Zannier Hotels Sonop
© Alex Teuscher for Zannier
Hotels Sonop

p.16-23
Desert Whisper
© Gondwana Collection Namibia

p.24-31
Wolwedans Collection
p.24-29: © Wolwedans Collection
p.30-31: © Mlenny (iStock)

p.32-35
Little Kulala
© Teagan Cunniffe

p.36-41
Okahirongo Elephant Lodge
p.37, p.38-39, p.41 bottom:
© Sanctuary Retreats
p.40, p.41 top: © David Rogers

p.42-45
Scarabeo Camp
© Sophia van Den Hoek

p.46-51
Shipwreck Lodge
© Natural Selection

p.52-59
Dar Ahlam
© Nicolas Matheus for Dar Ahlam

p.60-65
Umnya Desert Camp
p.60: © Umnya Desert Camp,
© All About Wanderlust
p.62-65: © Umnya Desert Camp

p.66-69
!Xaus Lodge
© !Xaus Lodge

p.70-73
Sanbona
p.71, p.73:
© Sanbona Wildlife Reserve
p.72: © brytta (iStock)

p.74-79
Adrère Amellal
p.74-77: © Karim Moustafa/
Unique Captures Photography
p.78, p.79 top:
© Sameh Shahien Photography
p.79 bottom: **Faris Hamdan**

p.80-85
San Camp
© Natural Selection

ASIA, MIDDLE EAST AND EUROPE
© Six Senses Shaharut;
© ASSAF PINVHUK

p.88-93
Telal Resort Al Ain
p.89, p.90-91, p.93:
© Telal Resort Al Ain
p.92: © Nedal Haroon

p.94-99
Habitas AlUla
© Kleinjan Groenewald

p.100-109
Six Senses Shaharut
p.108, p.109: © Six Senses Shaharut:
p.102-103, p.104-105, p.107 top:
© Six Senses Shaharut;
© ASSAF PINVHUK
p.107 bottom: © Six Senses
Shaharut; © Amit Geron

p.110-113
Al Maha Desert Restort and Spa
© Courtesy Al Maha Luxury
Collection Resort and Spa

p.114-117
Three Camel Lodge
p.115, p.117 top:
© Three Camel Lodge
p.116, p.117 bottom:
© Ken Spence Photography

p.118-123
Suján the Serai
© SUJÁN

p.124-127
Kaner Retreat
© Himanshu Lakhwani

p.128-131
Thousand Nights
p.129, p.130, p.131 bottom:
© Thousand Nights Camp Oman
p.131 top: rchphoto (iStock)

Words and Concept: Karen Gardiner
Design: Carolina Amell
Editing: Léa Teuscher

Sign up for our newsletter with news about new and forthcoming publications on art, interior design, food & travel, photography and fashion as well as exclusive offers and events. If you have any questions or comments about the material in this book, please do not hesitate to contact our editorial team: **art@lannoo.com**

©**Lannoo Publishers**, Belgium, 2023
D/2023/45/20 – NUR 500 / 450
ISBN 978 94 014 8870 9
www.lannoo.com